D1160880

F. Heide   F. Wlotzka

Meteorites
Messengers from Space

With 111 Figures and 22 Tables

Springer-Verlag

Berlin Heidelberg New York
London Tokyo Paris
Hong Kong Barcelona
Budapest

Prof. Dr. F. Heide†

Dr. F. Wlotzka
Max-Planck-Institut für Chemie
Saarstr. 23
55122 Mainz, Germany

Translator

Dr. R.S. Clarke, Jr.
Former Curator of Meteorites
National Museum of Natural History
Smithsonian Institution
Washington, DC, USA

Translated by R.S. Clarke, Jr. and F. Wlotzka

Cover illustration: Wolfgang Dobler
By courtesy of "Sterne und Weltraum".
Issue August 1994

ISBN 3-540-58105-7 Springer-Verlag Berlin Heidelberg New York

Library of Congress Cataloging-in-Publication Data. Heide, Fritz, 1891– [Kleine Mete-oritenkunde. English], Meteorites: messengers from space/Fritz Heide, Frank Wlotzka. p. cm. Includes bibliographical references and index. ISBN 3-540-58105-7 (Berlin:alk. paper). – ISBN 0-387-58105-7 (New York:alk. paper) 1. Meteorites. I. Wlotzka, Frank, 1931– II. Title. QB755.H43 1995 523.5′1–dc20 94-40204

© Springer-Verlag Berlin Heidelberg 1995
Printed in Germany

Typesetting: Macmillan India Ltd., Bangalore-25

SPIN: 10080183   31/3130/SPS – 5 4 3 2 1 0 – Printed on acid-free paper

Preface

The German edition, "Kleine Meteoritenkunde", which first appeared in 1934, was written by Professor Fritz Heide, one of the few mineralogists of that time working actively in meteorite research. It was then one of the few books available describing the properties of meteorites and the phenomena connected with their fall. A second edition appeared in 1957, shortly before Heide's death. In the following years, great progress was made in meteorite research, mainly due to the increasing interest in space research. The landings on the moon were especially stimulating. New analytical methods, developed for the lunar samples, were also applied to meteorites. The understanding of physical processes (radioactive decay, interaction of cosmic rays with meteoritic matter in space) led to new insights into the genesis and evolution of meteorites. It was found that primitive meteorites exist, which contain a record of the early history of the Solar System, when the planets were forming. Most recently, particles have been found in meteorites – *stardust* – which formed even earlier and beyond the Solar System. Thus, a new, greatly revised edition of the book became necessary, which appeared in German in 1988.

The present text is a translation of this last edition, adapted to an international readership and augmented by new results acquired in the last years. An attempt was made to retain a virtue of Heide's original work, namely, to appeal to the interested laymen and to students of mineralogy, astronomy and neighboring fields, as well as meteorite collectors and experts. Thus, this book describes meteorite falls, including the impact and cratering phenomena connected with giant meteorites, the mineral and chemical composition of meteorites, their classification, their age and origin, as well as their significance for the early evolution of our Solar System. It also describes some of the methods used to obtain these results. An appendix for the meteorite collector shows how to calculate trade values for different meteorites, and a list of meteorite falls from the last 8 years is given.

Mainz, January 1995 Frank Wlotzka

Contents

Introduction

Certainly every reader of this small book has already seen a shooting star. Out of the clear night sky a small point of light, no larger than its fixed-star neighbors, silently shoots across a segment of the heavens and disappears just as suddenly and as quietly as it appeared. "A star has fallen from heaven," we say, and some take it as a good omen. Many readers, however, will have seen a rarer occurence in the night sky. Not just a point of light the size of a star, but a great ball accompanied by an eye-catching light display, radiant blue-white or reddish-yellow, silently traverses a great distance in the heavens and disappears below the horizon or suddenly extinguishes. These events are called *Meteors*. Still rarer and only witnessed by a few fortunate individuals, is a third, more dramatic phenomenon. The night suddenly becomes as bright as day. A large fireball leading a long, luminous trail is seen crossing the sky, and the sounds of whizzing and the clatter of thunder are heard. Lasting only a few spellbinding seconds, the display concludes with an explosive detonation. So intense is the event that it is completely apparent in the brightest daylight. People positioned at the end of the path may see solid bodies fall and penetrate the ground. When these are retrieved from the soil, they are found to be fragments with stony or metallic properties.

All three of these heavenly displays, so different in appearance, had the same origin: the collision of a solid body from space with our Earth. But only those bodies that are accompanied by the most spectacular displays penetrate the Earth's protective shield, the atmosphere, and fall to the Earth's surface as *Meteorites*. These meteorite specimens are exceedingly important to us. Before the return of lunar samples, they were the only materials from beyond the Earth that we could take in hand and study. The sun, the planets, and the stars, as well as shooting stars and meteors send messages to us only as visible and invisible radiation. To be sure, ingenious instrumentation has been developed that permits the interpretation of this radiation by indirect means in the attempt to understand the composition of these heavenly bodies, but these investigations leave many open questions. Meteorites, on the other hand, these palpable messengers from space, can be investigated using all the mineralogical, chemical, and physical methods with which the components of the Earth's solid crust are studied. We are already in the position, therefore, to answer satisfactorily a substantial number of the questions connected with the existence of these messengers from space.

The consequences of these investigations go well beyond the realm of pure science. A knowledge of the composition of the components of meteorites allows us to draw important conclusions about the development and composition of our own planet and to recognize principles that are of great importance for an understanding of the distribution of materials within our Earth. Such principles, for instance, regulate the development of productive mineral deposits that have great importance for man's cultural and economic activities and might otherwise not be understood.

Futhermore, meteorites can have a direct influence on the lives of individuals. So far we have mentioned only the ordinary cosmic projectiles for which the Earth serves as the endpoint of their journey through space, projectiles that inflict no serious damage. We know, however, that an occasional visitor from space can be of great size and that its arrival on Earth is accompanied by gigantic explosions.

In this little book we will focus on the questions facing one who takes in hand such remarkable creations for study. Shooting stars and meteors that glow in heavens' heights and are known only from the light they emit will be dealt with only insofar as they relate to meteorites.

The term *meteorite* is derived from the Greek. It means *that which originates in the atmosphere*. Today, we call a *meteoroid* a small object traveling through space; a *meteor* its luminous trail in the sky; it becomes a *meteorite*, when it reaches the Earth's surface.

1 Fall Phenomena

1.1 Light Phenomena

The first indications that a meteorite gives of its arrival on Earth are the light and sound phenomena that accompany its fall. The large distances over which these phenomena are visible and audible, and the resulting excitement produced, have induced people to promptly set down their observations in writing. As several hundred meteorite falls have been observed in the last century, the reports of eyewitnesses comprise voluminous literature.

On the other hand, the existing pictoral records are less satisfactory. As meteorites arrive unannounced, photographing meteorite falls during daylight hours has still been unsuccessful. Only sketches and paintings that were prepared later based on the reports of eye witnesses are available. Albrecht Dürer, the famous painter, painted the explosion of the Ensisheim fireball from his own observations. This painting has only recently been detected on the back of another Dürer painting. It was reproduced in *Meteoritics* in an article by U. Marvin (1992) on the Ensisheim fall. Figure 1 shows a sketch of the meteorite fall from Ochansk,

Fig. 1. Fall of the Ochansk meteorite, Russia, August 30, 1887, 12:30 h. (After Farrington, Meteorites, 1915)

near Perm, Russia, on 30 August 1887, at 12:30 p.m. A fiery body appeared in the heavens, leaving a luminous smoke cloud streak behind, and gliding with only a slightly inclined path through the sky. This spectacle was seen for only 2–3 seconds, and after 2–3 minutes loud reports were heard as if numerous cannons had been fired. The largest meteorite whose fall was observed, from Sikhote-Alin in the far east of Russia, north of Vladivostok, displayed very similar but more intense fall characteristics (Fig. 2). A blindingly bright fireball shot across the sunny sky for a few seconds. The brightness of the fireball was so great that it hurt the eyes, and behind it an enormous smoke trail remained that could still be seen a few hours later. Soon after the disappearance of the fireball powerful thunderclaps and explosive noises could be heard.

Fig. 2. Fall of the Sikhote-Alin iron meteorite north of Wladiwostok, Sibiria, February 12, 1947, about 10 : 30 h. (After a drawing by P.J. Medwedew, Committee on Meteorites, Moscow)

The intensity of the light occurring by a daylight fall is commonly compared with that of the Sun. At night vast regions become as bright as day and one could easily read a newspaper. The luminous trail that a falling meteorite inscribes in the heavens may be observed over very great distances. The Prambachkirchen

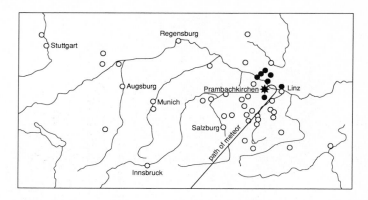

Fig. 3. Path of the Prambachkirchen meteorite fall, November 5, 1932. *Open circles* Sightings of the fireball. *Filled circles* Sound also heard. (After J. Rosenhagen, Jahrbuch Oberöster. Musealverein 86, 1935)

meteorite that fell near Linz, Austria, on the night of 5 November 1932, bathed large areas of southern Germany and northern Austria in bright light for several seconds. It was seen in Stuttgart, Regensburg, Innsbruck, Munich, and many localities in the Linz region (Fig. 3).

The beginning of the luminous path is usually difficult to determine. It is only by chance that an observer might look in that region of the heavens from which the meteorite appears. Most are attracted by the appearance of the bright light. The actual height of this first light appearance can be calculated from the testimony of several eyewitnesses, as is described in the following paragraph. By this means a beginning height of approximately 300 km was obtained for the path of the Pultusk, Poland, stony meteorite fall of 1868, and one of about 80 km for the Treysa, Germany, fall of 1916 (see also Table 1).

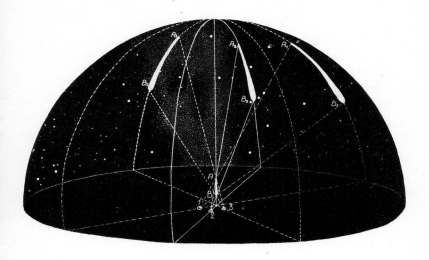

Fig. 4. Real and apparent paths of a meteorite

Each observer of a meteor will see its path in a different region of the sky, depending on the position of the observer. From a number of such *apparent paths*, however, the true trajectory may be calculated. With the aid of a sketch (Fig. 4), these relationships can be made clear. The small bright path AB is the actual path of a meteorite that is racing perpendicularly to Earth. The small circles 1,2,3 indicate the locations of three different observers. They are encircled by the bell-shaped hemisphere of the diagram as a common firmament. The horizontal circle on which the bell sits represents the horizon. On the rear inner wall of the hemisphere the Milky Way and some constellations are indicated. Next to and to the right of the Milky Way at B2, one recognizes Orion with its belt of three stars. The real path of the meteorite in relation to the very far distant heavens is found to be only a comparatively small distance from the observer, and it is seen projected in very different locations in the sky. Observer 1 sees the path to the right in the heavens in the direction from A1 to B1. The image lies, of course, on the sectional plane through the hollow sphere which is determined by the two sighting rays 1-A-A1 and 1-B-B1. The plane of section is indicated in the diagram with the same label as the sighting ray. Observer 2, on the other hand, sees the meteorite shoot through the middle of Orion right to the three belt stars: A2-B2. Observer 3 finally sees the apparent path traverse the Milky Way diagonally from A3 to B3. In Fig. 4 we see that the three sectional planes containing the apparent paths intersect each other in the line AB. This line of intersection is the true path. In practice the calculation of an accurate meteorite trajectory is naturally considerably more difficult. With such a sudden occurrence as a meteorite fall, it is only possible to establish the apparent paths with large inaccuracies.

The calculation of a meteorite trajectory is possible with greater accuracy if it is photographed from several locations. This idea has led to the establishment of several networks of automatic stations that continually photograph the night sky. These have been developed in Czechoslovakia, the USA, and Canada. The Prairie Network in the USA (which is no longer operative) had 16 stations in the Midwest between Oklahoma and South Dakota distributed over an area of 1000×1000 km. Each had four cameras directed toward the four quadrants of the sky (Fig. 5). Using a rotating shutter, the meteor trail was interrupted 20 times per second, allowing the calculation of the velocity of the body from the length of the segments. Naturally, one also wanted to find the fallen meteorite

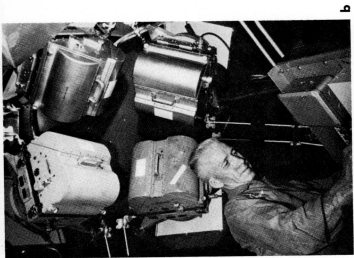

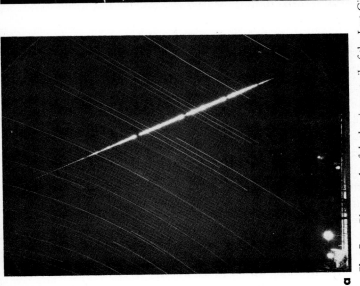

Fig. 5. a Photograph of the luminous trail of the Lost City meteorite by an automatic camera station of the Prairie Network. In the background trails of stars recorded during the long exposure time before the fall. **b** Inside the automatic camera station: the four cameras with the film magazines. (Photographs Smithsonian Astrophysical Observatory, Sky and Telescope 39, 1970)

using the calculated, true fall path. All three networks were successful in recovering a meteorite: Pribram in Czechoslovakia (fall of 7 April 1959), Lost City in the USA (3 January 1970), and Innisfree in Cananda (5 February 1977). All three are stony meteorites. The special importance of these recoveries is that the accurately measured, luminous paths allow the reconstruction of the orbits of these meteorites around the Sun. Thus, information about their origin (see Sect. 3.2) can be obtained.

At the end of the visible path explosive episodes are frequently observed, whereby the meteorite bursts into several parts. This point is called the *retardation point*, its importance will be elucidated further in Section 1.6.1. The height of the end of the path depends upon the initial velocity and the mass of the meteorite. Determined values from path observations lie between 4 and 42 km. In Table 1 the beginning point, the endpoint, and the initial velocity of the three photographically recorded meteorite paths are presented.

The course of the path is generally drawn as a straight or only slightly curved line, but larger deviations can occur. The meteorite from Prambachkirchen in northern Austria (fall of 5 November 1932) burst at a height of 14 km into two parts, and the smaller fragment that had a recovery weight of 2 kg traversed an arc of 10 km radius. It thereby changed its direction of flight by more than $180°$ (Fig. 3).

The color of the light is varied, mostly white is observed, but also green, red, and yellow are seen. It is not necessarily the same for all parts of the path. This change in color cannot be connected with a change in composition of the atmosphere with height, as we know from other observations (for example,

Table 1. Meteorite paths in the atmosphere

Name	Type	Initial weight (kg)	Final weight (kg)	Beginning of luminous path (km)	End of luminous path (km)	Initial velocity (km/s)
Pribram	H-chondrite	1300	53[a]	100	12	21
Lost City	H-chondrite	50	17	90	19	14
Innisfree	LL-chondrite	15	5	70	20	14.5

Source: Revelle and Rajan, J Geophys Res 84 (1979).

[a] Estimated weight, only 6 kg was found.

the northern lights) that nitrogen and oxygen comprise the major atmospheric components up to the greatest heights.

Sudden light bursts, often multiples, are sometimes observed. Figure 6 reproduces such a light burst of a bright meteor. The light emanates from a mainly round or pear-shaped, glowing cloud of gas. This appears to the observer much larger than the meteorite itself. Where estimates of the diameter of fireballs based on eye witness reports were possible, values of several hundred meters always resulted; for example, 300 m for the meteorite from Pultusk. The greatest brightness of the Treysa fireball, at an altitude of 50 km, gave a diameter of 1000 m, which was reduced to about 400 m further along its path. Actually, the diameter of the Treysa meteorite was only 36 cm.

The gas cloud surrounding the meteorite is the result of its intense volatilization. Entering the atmosphere at high velocities of 15 to 70 km/s, the surface of the meteorite collides with air molecules and is intensely heated, causing a thin layer of the surface to melt and vaporize. The resulting gas cloud, the *coma*, is incited through further collisions and becomes luminescent. A train of ionized air molecules follows behind and glows feebly. This ionization trail can be recorded by radar as it reflects radar waves. Thus, radar provides another method for observing meteors in the high atmosphere.

For meteorites that fall during the day a smoke trail can often also be seen. The smoke trail accompanying the fall of the Sikhote-Alin meteorite (Fig. 2) remained for several hours; it was so thick that the Sun behind it appeared as a weakly luminous red disk. The smoke consisted of very fine particles of

Fig. 6. Multiple light bursts of a meteor on July 26, 1952, 00 : 08 h.
(Photograph A. Ahnert, Sonneberg Observatory, Thüringen)

meteorite material that originated from the atomization of the incandescence melt coating the meteorite. A substantial part of the original mass of a meteorite is thus removed by vaporization and dust production. Table 1 shows that the stony meteorites Lost City and Innisfree lost two-thirds of their mass by atmospheric burning, while Pribram lost a surprising 96%.

1.2 Sound Phenomena

Still more impressive and more terrifying than the re-markable light phenomena are the sound phenomena that are associated with meteorite falls. They are so terrifying that people have fallen down from fright or have immediately panicked and sought cover in build-ings or under trees. Depending on the location of ob-servers, there is a whole scale of sounds that may be experienced, from thunder-like claps that rattle win-dows, to cannon detonations and small weapons fire, or to the sounds of roaring trains. Sounds reminis-cent of stormy weather and hissing sounds are often reported.

The manifestations of sound related to a me-teorite fall may be heard over a large area. Where investigations have been conducted, reports of fall-related sounds have come from areas extending over a 60–70 km radius. Similar to actual cannon fire, more than one zone of sound has been occasionally re-ported. In the case of the Treysa fall mentioned ear-lier, the inner sound area had a radius of about 60 km, however, from a distance of 95 to 120 km, sound phenomena could again be registered (Fig. 7).

The origin of the sound phenomena has been demonstrated experimentally. The factors that cause the sounds of meteorites are probably the same as those for high velocity projectiles. Figure 8 shows a very fast light artillery projectile during flight. We can see clearly that a cone-shaped sound wave ra-diates from the point of the projectile, the so-called shock wave, which produces the sound of thunder. The turbulent air behind the projectile as well as

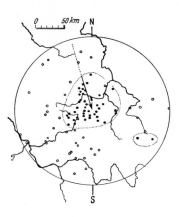

Fig. 7. Area where the Treysa meteorite fall (Hessen, Germany, April 3, 1916) was seen and heard (*filled circles*), or only seen (*open circles*). (After A. Wegener, Schriften der Gesell. z. Beförd. d. ges. Naturwiss., Marburg, 1917)

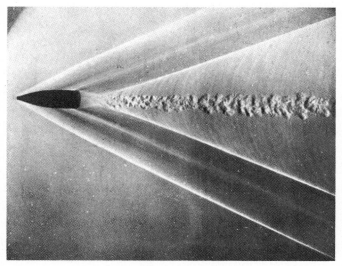

Fig. 8. Front wave of an infantry bullet. (After P.P. Ewald, Kristalle und Röntgenstrahlen, 1927)

reflections of the sound waves on clouds and the ground produce the more rolling types of sounds. The machine gun clatter heard by many observers is caused by the breaking of small fragments from the main mass, each one of which then causes its own shock wave.

1.3 The Main Meteorite Groups

Light and sound phenomena are similar for all types of meteorites. Before we discuss the frequency of meteorite falls and the consequences of their fall, however, an introduction to the different meteorite types should be given (see also Chap. 2). The most striking difference is between the *iron meteorites* that are composed of 90% or more metal and the *stony meteorites*, which consist essentially of rocky material. The stony meteorites are divided into the *chondrites* and *achondrites*. Chondrites consist of up to 80% approximately mm–size silicate spheres called *chondrules* (from the Greek *chondros* meaning grain or kernel) that are embedded in a fine-grained matrix. The achondrites lack these spherical structures (Greek prefix *a*- meaning absent) and are thus more similar to terrestrial igneous rocks. The chondrites also contain grains of metallic nickel-iron, but the silicates are always the major component. In addition, there is the much rarer group of *stony iron meteorites* that are composed of about equal parts of stony material and metal. These are not just a simple mixture of the first two types but represent a third distinct group. Stony irons are further divided into two classes, the *pallasites* and the *mesosiderites*. Characteristics which distinguish the groups are given in Table 2.

Table 2. Main meteorite groups

Group	Main characteristic	Main components
Iron meteorites	More than 90% metal	Nickel–iron
Stone meteorites:	More than 75% stony material	
Chondrites	Chondrules	Silicates, 3 to 23% metal
Achondrites	No chondrules	Silicates
Stony iron meteorites:		
Pallasites	Cm-sized silicate crystals in metal	Olivine, metal ca. 2:1
Mesosiderites	Silicate and metal intimately intergrown	Silicate, metal ca. 1:1

1.4 Effects on Landing

As conspicuous and impressive as the light and sound phenomena are to the observer of a meteorite fall, the effects actually produced on the Earth's solid surface are astonishingly modest. One of the largest known stones (observed fall 18 February 1948) is the 1-ton stone from Norton County, Kansas. It penetrated to a depth of only 3 m. How unobtrusive the landing place of small meteorites is, is shown in Fig. 9. The St. Michel stone meteorite that struck here weighed 7 kg. The holes are mainly roundish and strike either straight down or somewhat diagonally, as shown in Fig. 10 for the entry channel of the 2-kg Prambachkirchen stone meteorite. Similarly unremarkable are the impact pits caused by the frequently much heavier iron meteorites. The largest intact iron meteorite from the Sikhote-Alin meteorite shower from

Fig. 9. Impact pit of the stony meteorite St. Michel, Finland, July 12, 1910. (After Borgström, Bull. de la Comm. Géolog. de Finlande, 34)

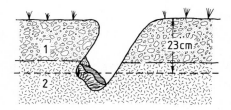

Fig. 10. Impact channel and situation of the stony meteorite Prambachkirchen. *1* Loose top soil; *2* loam. (After J. Schadler, Jahrbuch Oberöster. Musealverein 86, 1935)

eastern Siberia, 1.75 t in weight, was found in a small crater 4 m deep. The larger projectiles of the shower, which produced craters up to 26.5 m in diameter, were completely broken up in the process (Fig. 11).

These reports of only slight damage, which can easily be supported by many more historic examples, are in many respects rather remarkable. The condition of the ground, of course, would naturally be expected to have an influence on the amount of damage that is left behind. The harder the ground, the slighter the effects. A meteorite falling on hard rock, particularly

Fig. 11. Inside slope of the largest crater from the Sikhote–Alin iron meteorite shower, February 12, 1947; diameter 26.5 m, depth 6 m. (Photograph by A. Krinov)

a stony meteorite, would be completely shattered and would leave only slight evidence of the fall behind.

Particularly astonishing is the fact that, on the one hand, the effects on landing are very slight, and that, on the other hand, we have noted that meteorites enter our atmosphere with velocities of many km/s, velocities that far surpass our fastest projectiles on Earth. In fact, the effects left behind are less than those produced by duds from an intermediate caliber canon. Consideration of a larger body of material reveals that for the same ground conditions the intensity of the effects apparently depends only on the mass of the meteorite. Equal masses will cause about the same damage, despite the fact that their entrance velocities can differ greatly, ranging somewhere between 15 and 70 km/s. The reason for this peculiarity lies in the presence of our atmosphere. Despite its gaseous nature, its presence provides a protective shield from projectiles from space.

We all are familiar with the resistance of air, particularly when we experience higher velocities, such as traveling on a motorcyle. It depends on the cross-sectional area of a moving body, but more importantly on the velocity. Beyond the velocity of sound (330 m/s) the air resistance increases with the square of the velocity, i.e., it increases by a factor of 4 with a doubling of the velocity. Meteorites with higher entrance velocities, therefore, will be more strongly decelerated in the atmosphere than ones with smaller velocities. The atmosphere then has an equalizing influence on the different entrance velocities of meteorites.

Figure 12 presents a plot of the velocity and the height of the photographically determined path of the Innisfree meteorite. The entrance velocity of 14.5 km/s was essentially constant for the first seconds as the meteorite descended from 70 to 35 km, however, it was then rapidly reduced to 5 km/s in the following 15 km. At a height of about 20 km the meteorite had reached its previously mentioned *retardation point*. Its cosmic velocity has been overcome, and the meteorite continues to fall only under the influence of its weight and the resistance of air. The uniform end velocity with which the meteorite descends is, when compared to its entrance velocity, exceedingly small. It is so small, in fact, that the meteorite no longer gives off a luminous gas cloud and friction no longer melts its surface. Light phenomena cease at the retardation point. Observers who find themselves near the landing point of the meteorite see it fall from the heavens as a dark body.

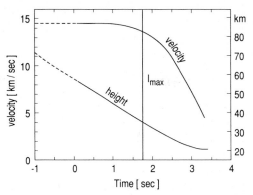

Fig. 12. Velocity and height of the Innisfree meteorite fall, calculated from the trail recorded by the Canadian Camera Network. At I_{max} highest light intensity. (After Halliday, Griffin and Blackwell, Meteoritics 16, 1981)

It is necessary, however, to qualify these considerations. They are appropriate only for masses of a size that are retarded by our atmosphere. Extremely large meteorites change the quantitative relationships discussed above in important ways, because the Earth's atmosphere is no longer able to completely overcome their cosmic velocity before they strike the Earth's surface. This will be discussed further in Section 1.6.

There is a another way in which meteorites behave differently than one might expect based on their dramatic manifestations in the atmosphere. As mentioned earlier, the meteorite surfaces are heated by collisions with air molecules and become molten. During this process, however, the main body of the meteorite is not significantly heated. This has been demonstrated directly: many stony meteorites that were recovered immediately upon landing were at most lukewarm. A few of them had even fallen into haystacks or into barns without igniting easily flammable materials. Their recovery sites also showed no indication of being burned, although the meteorite which fell at Alfianello, Italy (16 Feb. 1883) was reported to have somewhat singed the grass. Iron meteorites have occasionally been reported to be still hot. An example is the iron meteorite from Braunau, Czechoslovakia (14 July 1847) that was still too hot to be picked up 6 h after the fall. Another fragment from this same meteorite, nevertheless, hit the thatched loft of a house without noticeable singeing. Up to now there has never been a fire caused by an iron meteorite. By investigating the interior structure of meteorites one can determine that most of them could not have been strongly heated since their formation, certainly not to melting temperatures. Evidence for heating is found only in the few millimeters thick *heat-affected zone* directly under the fusion crust. This is due to the fact that the heating is very rapid and lasts only a short time. Heat requires time to penetrate into the interior, and the heating period lasts only a few seconds. The thin melt layer on the surface cools rapidly during final descent and solidifies to a firm crust well before the meteorite reaches the Earth.

1.5 Meteorite Showers

We have seen that when a small meteorite enters our atmosphere little damage occurs on landing. When a much larger meteorite enters the atmosphere, damage

Table 3. Meteorite showers

Name and date	Fall area (km)	Class	Number of pieces	Total weight
L'Aigle, France 1803 April 26	2 x 4	L-chondrite	~ 3000	~ 40 kg
Stannern, CSR 1808 May 22	13 x 4.5	Eucrite	200-300	~ 52 kg
Knyahinya, Ukraine 1866 June 9	14.5 x 4.5	L-chondrite	> 1000	~ 500 kg
Pultusk, Poland 1868 January 30	8 x 1.5	H-chondrite	~ 100 000	~ 2 t
Hessle, Sweden 1869 January 1	16 x 4.5	H-chondrite	Many	~ 23 kg
Khairpur, Pakistan 1873 September 23	25 x 4.5	E-chondrite	Many	> 15 kg
Homestead, USA 1875 February 12	10 x 5	L-chondrite	> 100	230 kg
Mocs, Romania 1882 February 3	14.5 x 3	L-chondrite	> 3000	300 kg
Holbrook, USA 1912 July 19	4.5 x 0.9	H-chondrite	~ 1400	218 kg
Sikhote-Alin, Sibiria 1947 February 12	2.1 x 1	Octahedrite	Many	~ 70 t
Allende, Mexico 1969 February 8	50 x 12	CV-chondrite	several thousand	~ 2 t
Jilin, China 1976 March 8	72 x 8.5	H-chondrite	> 200	~ 4 t
Mbale, Uganda 1992 August 14	8 x 5	L-chondrite	> 50	> 100 kg

is still generally surprisingly light. Meteorite *showers*, formed by the breakup of large bodies in the upper atmosphere, occur frequently among meteorite falls, producing many small fragments that individually do little damage. The individual pieces can number into the thousands, e.g., the Pultusk (1868) fall was estimated to consist of 100000 fragments, many of which were only pea size. Most showers have been composed of stony meteorites, but one iron meteorite shower has been observed, the Sikhote-Alin fall of 1947. Sometimes many iron meteorite fragments are found distributed in the same general area, leading us to assume that they fell as showers. Examples are the Gibeon, Namibia, finds (more than 50 pieces), and the Toluca and Coahuila finds in Mexico. Table 3 lists some of the largest meteorite showers.

The meteorites in a shower are frequently distributed over an elliptic area, as illustrated in Fig. 13 by the Homestead shower. The direction of shower movement in the atmosphere was from south to north. Normally, the largest pieces in a shower travel the farthest due to their higher kinetic energy; correspondingly, the largest pieces are found at the north end of the Homestead ellipse, the heaviest among them weighing 32 kg.

Meteorite showers originate in the breakup of a larger body entering the atmosphere. This apparently occurs at high altitudes, so that each individual piece develops its own fusion crust. Sometimes one finds thinner, secondary crusts on certain faces of a stone showing that a secondary breakup took place later. This fragmentation has also been recorded in photographs of meteorite paths. The path of the Innisfree meteorite divided into six separate luminous trails at

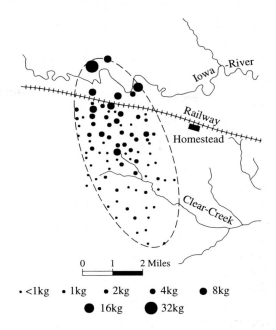

Fig. 13. Fall ellipse of the Homestead, Iowa (USA), meteorite shower, on February 12, 1875. (After Farrington)

heights between 30 and 25 km. All six fragments were later found within an area of a few hundred meters.

One of the largest meteorite showers in recent times was the Allende, Mexico, fall. Early in the morning of 8 February 1969 a bright fireball appeared over northern Mexico and thousands of stones fell over an area of 300 km². The ellipse of the fall was 50 km long and 12 km wide, the largest stone of 110 kg was found at its northern tip (Fig. 14). It was estimated that the weight of the recovered specimens was 2 t and that only half the material was recovered (Fig. 15). The Allende fall was of particular importance for science.

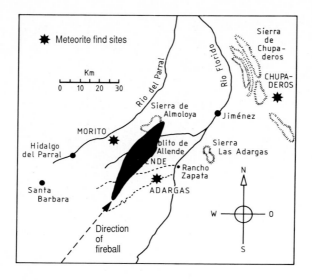

Fig. 14. Fall ellipse of the Allende, Chihuahua, Mexico, meteorite shower. Three large iron meteorites were found in the same area: Morito, 11 t, found about 1600; Adargas, 3.4 t, known since 1600; Chupaderos, 14 and 6.7 t, found 1852. (After Clarke et al., Smithson. Contr. Earth Sci. 5, 1970)

The recovered stones belonged to a relatively rare type of stony meteorite, the carbonaceous chondrites. This material is more pristine than other meteorite types; it comes close to the unaltered original material from which the Solar System originated. Moreover, the first lunar rocks were brought to Earth in 1969, and the same laboratories that had developed new and especially sensitive research techniques to study them, suddenly also had abundant material available from Allende for investigation. (The total weight of the 36 carbonaceous chondrites known before 1969 came to only 420 kg, but specimens were distributed in many

Fig. 15. An individual from the Allende meteorite shower, as it was found in the field. (After Clarke et al., Smithson. Contr. Earth Sci. 5,1970)

and often inaccessible collections.) Much of today's understanding of our Solar System and its origins was derived from Allende material. Especially the realization that its white calcium- and aluminium-rich inclusions may be direct condensates from the Solar Nebula led to the thorough investigation of such materials for the first time (see Sect. 3.3).

1.6 Hypervelocity Impacts

1.6.1 Meteorite Craters

The material presented so far has demonstrated that meteorites of known size cause astonishingly little damage on landing. On the other hand, a number of pronounced crater-like structures not connected to volcanoes have been found on the surface of the Earth. These structures are very similar to each other, and some of them have been thoroughly investigated. These investigations concluded that many of them can only be the consequence of large meteorite impacts. These impact sites are of such magnitude that we must set aside completely the simple picture for the fall of small meteorites if we are to understand what took place. Their enormous extent required meteorite impacts of gigantic force, far beyond anything previously recognized on Earth. The question of their origin has evolved in recent years from a minor speciality within meteoritics to a highly interdisciplinary field of universal significance for science. Should a giant meteorite fall into a heavily populated area, for instance, it would result in a tragedy of unimaginable magnitude. If the target was one of the major cities of the world, perhaps Paris, London or New York, probably nothing of the city or its residents would remain.

Because only one such crater was recognized for many years, Meteor Crater, Arizona, serious doubts were expressed regarding its meteoritic origin. In the past years a large number of these craters have been discovered and accepted with certainty, some with reservations, as the impact sites of giant meteorites. A number of years ago nature kindly provided us with an opportunity to observe the consequences of the fall of such a giant meteorite, the Tunguska event in Siberia, described below. This fall established beyond any doubt that meteorites of much greater size and with velocities not previously experienced do occasionally strike the Earth.

The great size and the high velocity are related, because the Earth's atmosphere is not capable of

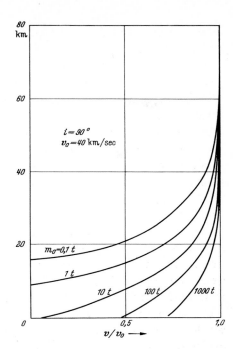

Fig. 16. Deceleration of meteorites with different mass

retarding such a large mass. As seen in Fig. 16, with increasing mass the retardation point penetrates progressively deeper into the atmosphere, finally reaching the surface of the Earth. In other words, these giant meteorites strike the surface of the earth after having lost only a small fraction of their cosmic velocity. How high the impact velocity may become is indicated in Fig. 16. The meteorites considered range in mass, m_0, from 0.1 to 1000 t, the approach is perpendicular ($i = 90°$), and the entrance velocity, v_0, is 40 km/s. The curves show the decreasing velocity as the fraction v/v_0 of the entrance velocity. One sees that the masses from 0.1 to 1 t are retarded between an altitude of 16 to 8 km. Under these conditions a meteorite of 10 t will be almost completely retarded at the Earth's surface, while one of about 100 t strikes at 20 km/s and one of 1000 t at 29 km/s. The kinetic energy is $E = 1/2mv^2$ (m = mass, v = velocity), and this combination of increased mass and velocity produces vast amounts of energy. The unit of energy is the joule (J), and 1 joule = $1 \text{ kg} \cdot (\text{m/s})^2$. The result is that for an iron meteorite of 100 t impacting the earth with a velocity of 20 km/s, the energy is 2×10^{13} J; for an iron meteorite of 100 t striking at 29 km/s, it is 4×10^{14} J. To produce a crater 1200 m in diameter, such as the Meteor Crater at Canyon Diablo, Arizona (described below), much greater energy is necessary. One can estimate from model calculations that this crater was produced by an iron meteorite of 100000 t with an entrance velocity of 15 km/s, and with a corresponding kinetic energy of 1.12×10^{16} J or about 1 megaton of TNT.

The great kinetic energy released by such a body striking the surface of the Earth produces completely different effects than the free fall of a small meteorite that has been retarded in the atmosphere. The energy is largely converted to heat that is so great that the meteorite itself is explosively vaporized and the rock at the impact site is partly vaporized and melted. The resulting crater is fundamentally an explosion crater and is thus different from the simple penetration holes considered above. The great kinetic energy also generates a hypervelocity shock wave that modifies the rocks surrounding the impact site, thus leaving a permanent record of the impact. Conversely, the finding of such shock transformations present evidence that a particular area must have been the site of a giant meteorite

Fig. 17. World map with known impact craters, as of 1987. *Open symbols* probable meteorite craters; *closed symbols* Proven meteorite craters. (After Grieve, Geol. Soc. Am. Spec. Pap. 190, 1982, and Grieve and Robertson, Geol. Surv. Can. Map 1658A, 1987)

33

impact. These effects will be treated in more detail below. First, we will discuss historic developments and describe craters in the order of their discovery. The map in Fig. 17 shows the locations of impact craters known to date.

The longest known and most throughly studied meteorite crater is the Canyon Diablo or *Meteor Crater* in Arizona, USA. The lower half of Fig. 18 is an aerial photograph of this structure. The crater lies in a region of very flat desert formed on limestone, which is underlain at greater depths by white and red sandstones. Almost circular in outline, its largest diameter is 1186 m and the depth from the crater rim to the floor is now 167 m. Its dimensions become obvious in the upper half of Fig. 18, in which the city of Jena is superimposed, illustrating the approximate size of the crater. The more than 40-m-high rim is composed of fractured rock layers and loose materials which were ejected from the crater; among the debris are large blocks up to 4000 t. A section through the crater is shown in Fig. 19. The sandstone of the crater floor is completely pulverized to great depths, to the extent that it easily crumbles to powder in the hand. In part, it is also sintered and melted. At greater depths, however, the sandstone strata are completely undisturbed, a finding of great importance. The crater floor itself is covered by a thin layer of sediments that were deposited from a small post-impact lake that has since dried up.

The age of the crater is not known with certainty. A 700-year-old cedar tree that grew on its rim provides a minimum value. An age of 20000 to 30000 years has been estimated from the degree of weathering of the limestone. It is therefore questionable whether the fall of the giant meteorite was

Fig. 18. Aerial view of Meteor Crater (Canyon Diablo, Arizona, USA). The *upper picture* shows the outline of the crater superimposed over the town of Jena, Germany. (After a photograph by Carl Zeiss, Inc.)

experienced by the indigenous Indians, although it is mentioned in their legends.

Accordingly, one of their gods is supposed to have descended from heaven to this place, accompanied by thunder and lightening, to be buried there. Even today an Indian following tradition is forbidden to visit the crater, it is taboo, and it is significant that the Indians have not participated in the search for meteorites in the crater region.

But where are the remains of the giant projectile that excavated this gaping hole? In the area within a few miles of the crater, but not in the crater itself, a large number of iron meteorites have been

35

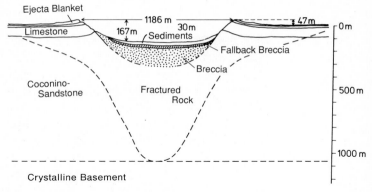

Fig. 19. Cross section through Meteor Crater (Canyon Diablo, Arizona, USA). (After Roddy, Proc. 11th Lunar Planet. Sci. Conf., 1980)

Fig. 20. A piece of the Canyon Diablo iron with iron shale. (After Merrill, U.S. Nat. Mus. Bull. 94)

found over the years, with an estimated total weight of more than 30 t. The meteorite specimens were found intermixed with terrestrial material that was ejected from the crater. Along with the nickel-iron masses, one finds also a large number of solid, rust-brown specimens that essentially consist of iron and nickel oxides mixed with small amounts of limy and other terrestrial materials. Because of its layered structure, early workers named this material *iron shale*. It is formed as a consequence of weathering, in the course of which the nickel-iron metal is dissolved and then precipitated in the immediate surroundings as oxide. In the specimen shown in Fig. 20 the nucleus is composed of massive nickel-iron and the surrounding rim of iron shale.

The cumulative weight of all of these meteoritic materials amounts to only a few tons, far short of the mass necessary to have produced this giant structure. At the beginning of this century and based on the findings of artillery experts, it was estimated that a projectile of about 150 m in diameter and weighing more than about 10 million tons would be required. If the bulk of this mass were present and buried under the crater, then not only millions of tons of massive iron and hundreds of thousands of tons of nickel, but also cobalt and platinum metals in major amounts would be found. Thus, it became not only a scientific but also a commercially important object of the highest interest. These commercial considerations gave rise to extensive prospecting of the crater as a possible mine. The mining engineer D.M. Barringer acquired the prospecting license in 1903, and by 1920 a total of 28 drill holes and prospecting shafts had been sunk into the crater floor. These led to an unusual, complete understanding of the crater's structure, but no progress was made toward finding the giant meteorite. Because of this disappointment, it was then assumed that the meteorite's path was diagonal from north to south and that it would not be found under the crater floor but under its southern wall. A new drill hole was sunk on the rim, and at depth some iron shale was found, but at 420 m the drill bit stuck. New financial backing was organized, and a shaft was courageously sunk south of the crater rim in order to find the presumed meteorite. At a depth of about 200 m the shaft flooded. At this point prospecting operations temporarily ended, the exploration of the crater having cost about $ 600,000 altogether. Throughout these setbacks Barringer continued to hold firmly to the theory of meteoritic origin, even in the face of the opposition of a number of then prominent geologists. In his honor, the crater is also referred to as Barringer Crater.

Although no projectile was found in the crater, it became increasingly clear throughout these prospecting operations that this crater must have originated by meteorite impact. The nature of the rocks (only sedimentary rocks and no trace of eruptive rocks were found), the undisturbed layering of these rocks at depth, the occurrence of sintered sandstone, and the mixing of abundant meteorite material with the debris in the crater walls rule out any volcanic structures, such as the Eifel Craters. Likewise, a sinkhole structure, or a collapsed salt dome at greater depth, or a crater formed by a natural gas explosion could all be ruled out on these grounds.

As the controversy over the nature of the Meteor Crater was at its height, a new structure of this type was discovered in 1928 near Odessa, Texas (USA). To be sure, it is much smaller (162 m diameter, 30 m depth), but otherwise very similar to the one in Arizona. Here, too, iron meteorites and a great deal of iron shale were found outside the crater.

Fig. 21. Irregular piece of iron from Henbury crater, Australia, half the natural size

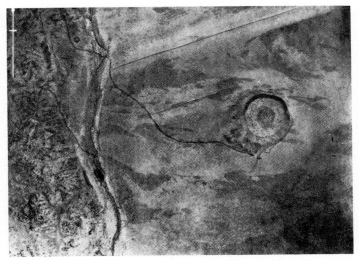

Fig. 22. Aerial view of the Wolf Creek meteorite crater, Western Australia, diameter 840 m. (After Guppy and Mathesun, J. Geol. 1950)

Further discoveries followed then in rapid succession. In 1930 R.A. Aldermann detected a crater field at Henbury in Australia (Northern Territories) with a total of 13 craters. Iron meteorite fragments were also found outside of the craters. These fragments often have interesting shapes, as if they had been torn from

39

Fig. 23. Aerial view of the main crater from Kaalijärv, Saaremaa Island, Estonia, diameter about 100 m. (After Reinwald, Natur Volk, 1940)

a larger body (Fig. 21). Additional crater sites have since been found in Australia: Box Hole, Dalgaranga, and Wolf Creek. The well-preserved Wolf Creek Crater (850 m diameter) is the second largest *proven* (see below) impact crater known so far (Fig. 22).

In Europe the Kaalijärv Crater on the Island Saaremaa (Gulf of Riga) has been known for some years. Through the discovery of associated meteoritic iron, I.A. Reinwald established in 1937 that it was a meteorite crater. The main crater (Fig. 23), having a diameter of 110 m and an initial depth of 22 m, contains a small lake. In the surrounding area there are eight more, smaller craters with diameters of 12 to 50 m.

The craters described here and several other similar ones are listed in Table 4. Along with these *proven* meteorite craters, i.e., where meteorite material was found in the vicinity, there are other, frequently very

Table 4. Proven impact craters (with associated meteorites)

Name	Number	Largest diameter (m)	Age (Years)	Meteorite class
Boxhole, N.T. Australia	1	175	5400	Octahedrite
Campo del Cielo, Argentina	20[a]	90	4000–5000	Octahedrite
Dalgaranga, W.A. Australia	1	21	–	Mesosiderite
Haviland, Kansas, USA	1	11	–	Pallasite (Brenham)
Henbury, N.T. Australia	14	150	5000	Octahedrite
Kaalijärv, Estonia	9	110	3500–3900	Octahedrite
Meteor Crater, Arizona, USA	1	1200	20 000–30 000	Octahedrite
Morasko, Poland	7	100	–	Octahedrite
Odessa, Texas, USA	3	162	50 000–100 000	Octahedrite
Sikhote–Alin, Russia	122[a]	26.5	Fall 1947	Octahedrite
Wabar, Saudi Arabia	2	97	6400	Octahedrite
Wolf Creek, W.A. Australia	1	850	Pliocene	Octahedrite

[a] Including impact pits.

large and very old structures, for which the meteoritic nature cannot be directly established. Here, the previously mentioned shock wave transformations can be considered. For their production, pressures of more than 10000 atm (1 atm = 1 bar) are necessary. These pressures are reached only during hypervelocity meteorite impacts and not during normal geologic processes within the Earths's crust, such as volcanic eruptions.

1.6.2 Shock Effects

At the site of a hypervelocity meteorite impact the target rocks are transformed by the shock wave. Table 5 lists the observable transformations, grouped into the shock metamorphism stages 0 to 5. The recorded *rest temperature* is the residual temperature which remains after passage of the shock wave.

In addition to these microscopically observable transformations, there is also a rock structure, the so-called *shatter cone*, which is observable in hand specimens. Shatter cones are centimeter-to decimeter-sized, cone-shaped structures that are especially well developed in limestones (Fig. 24). The apices of the cones point to the center of the impact structure.

During crater formation, the rocks that have been broken and transformed or melted by shock heating are mixed together and in part ejected from the crater and redeposited around it. This results in rock debris surrounding the crater. These *impact breccias*, made up of variegated glass and rock fragments, characterize an impact structure.

Figure 25 illustrates the formation of an impact crater. Three stages are distinguished:

Table 5. Stages of shock metamorphism

Stage	Pressure range (GPa)[a]	Post-shock temperature (°C)	Shock effects
0	5–10	–	Rock fracturing, kink bands in mica[b]
I	10–35	100	Planar elements in quartz, deformation lamellae in feldspar, partial isotropization in quartz and feldspar
II	35–45	300	Diaplectic glasses of quartz and feldspar, coesite and stishovite in quartz
III	45–55	900	Feldspar melt glasses with vesicles and schlieren, thermal decomposition of mafic minerals
IV	55–80	1700–3000	Complete melting of rocks and mixing of melts
V	Above 80	More than 3000	Vaporization

Source: D. Stöffler, J Geophys Res 76 (1971).
[a] GPa = Giga Pascal, 1 GPa = 10 kbar = 10000 atm.
[b] Can also be formed by tectonic pressure.

Fig. 24. Shatter cones in limestone from the Steinheim basin, Germany. (After v. Engelhardt, Naturwissenschaften 61, 1974)

compression stage, excavation stage, and subsequent crater modification stage.

a) A meteorite strikes a solid target at 15 km/s. The resulting pressure exceeds 10^6 atm (100 GPa).

b) Compression: The meteorite and part of the target rock at the impact site melt and vaporize. From the depth-of-burst point a cone of liquid and vapor is sprayed with high velocity.

c, d) Ejection: A shock wave propagates radially through the rock and accelerates the material initially toward the bottom, and through interaction with the free surface, also horizontally. As a result, the melted and shocked material is forced sideways into the crater walls and can finally be ejected. The crater rim bulges as a consequence. At depth the result of the shock wave is a zone of decreasing shock metamorphism (see Fig. 29). The maximum size of the primary crater is

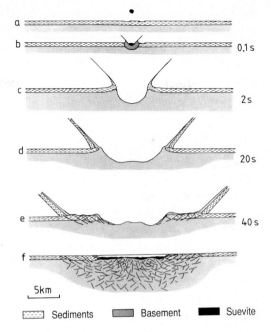

Fig. 25. Model of impact crater formation, see text. (After Pohl and Gall, Geol. Bavarica 76, 1977)

achieved by the ejection of material, however, some melted material remains in the crater.

e) Subsequent modifications: The primary crater is unstable; it is filled by material sliding from the walls into the crater, along with fallback of some of the ejected material, another part of which produces the ejecta blanket around the crater.

f) Final contour of the crater.

In the case of larger craters the relaxation and rebound after ejection from the primary crater can lead to the formation of ring structures and a central uplift.

Particularly impressive examples of such structures are the central mountains observed in large lunar craters. Remnants of central uplifts are also found in a number of terrestrial craters such as the Steinheim Basin and the Siljan structure.

1.6.3 The Ries Crater

One of the best studied of the large impact structures is the Ries Crater, Germany (Fig. 26), which includes within it the town of Nördlingen. The crater is a nearly circular, flat basin, 22 to 24 km in diameter, on the northern rim of Franconian–Swabian Alb. The crater rim and in its environs are covered by an ejecta blanket (Fig. 27). These ejecta consist mainly of what is called *Bunte Breccie*, an intricate mixture of various types of predominantly sedimentary rocks, and *suevite* which contains igneous and metamorphic rocks in all shock stages, and in addition, impact glass occurring in fragments or as flattened ejecta bombs, called *Flädle* (or Fladen), in a finely comminuted groundmass (Fig. 28). Earlier, these rocks were interpreted as volcanic tuff and the Ries structure as a volcanic crater. However, true volcanic lava such as basalt is absent, as is evidence for volcanic vents or processes. Instead, the breccia contains many indications of shock metamorphism, such as planar elements, diaplectic glass, and the high pressure modification of quartz, coesite. From modeling calculations it has been shown that a structure such as the Ries Crater could be formed by a stony meteorite body 600 to 2000 m in diameter striking at a velocity of 25 km/s. At a given diameter of the primary crater of 12 km, 200 to 2000 times the

Fig. 26. Aerial view of the Ries crater at Nördlingen, Germany. (Photograph A. Brugger, all rights Stadt Nördlingen)

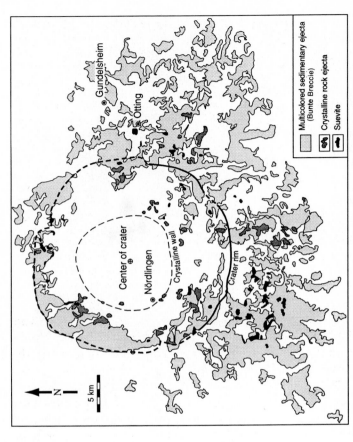

Fig. 27. Geologic map of the Ries, Germany, impact crater, showing how far Bunte Breccia and suevite were ejected from the crater. Ries ejecta to the north of the crater have already been removed by erosion. (After Chao, U.S. Geol. Survey Bull. 2050,

Fig. 28. Suevite from the Ries crater, Germany. In the middle a flattened, black glass bomb; coin diameter 2 cm. (Photograph G. Graup)

mass of the projectile would be excavated. From the distribution of the ejected material, the Bunte Breccie and the suevite, one can conclude that the body apparently penetrated 250 m of sedimentary strata and finally exploded in a crystalline basement at 1400 m. Figure 29 illustrates this model of origin of the Ries Crater in cross section. An estimated 3.5 km^3 of rocks was vaporized, 2 km^3 melted, and 130 km^3 ejected as far away as 40 km. Among the debris giant blocks of more than 100 m lenght can be found (Fig. 30).

The Steinheim Basin, a smaller crater, having a diameter of 3.4 km and an apparent depth of 90 m, lies 40 km southwest of the Ries Crater. In its center is a 50-m–high central uplift, called the Klosterberg, in which well-developed limestone shatter cones are found. Geological investigation revealed a 20- to 70-m-thick impact breccia layer in the crater. The

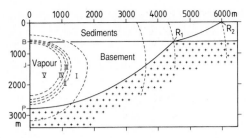

Fig. 29. Deep burst model for the formation of the Ries impact crater. The projectile penetrated 1400 m of rock before it exploded at *J*. Around the inner vapour sphere the rock was transformed into the metamorphic shock stages *IV* to *I*. All material above the line P-R_1-R_2 was ejected. (After v. Engelhardt and Graup, Geol. Rundsch. 73, 1984)

Fig. 30. Clod of White Jura at Ebermergen, which was transported about 10 km outward during the Ries crater event. (Photograph H. Gall, in: Das Nördlinger Ries, München 1983)

Steinheim Basin is the same age as the Ries Crater (15 million years); apparently, both structures were formed by the simultaneous impact of two projectiles.

1.6.4 Probable Meteorite Craters

Structures such as the Ries Crater and the Steinheim Basin are called *probable* meteorite craters to distinguish them from *proven* meteorite craters, around which meteorite material has been found. Today, we know of 12 *proven* meteorite craters and over 100 *probable* meteorite craters, the largest ones are listed in Table 6. When we examine the distribution of craters over the Earth's surface (Fig. 17), it becomes obvious that they are concentrated in certain regions such as Canada and northern Europe. These regions are so-called ancient cratons, regions that consist of geologically very old continental rocks that have not been covered with sediment layers nor have they been transformed by mountain building. In these regions craters are better preserved than elsewhere. In Europe, certainly, there is no crater better preserved than the relatively young Ries Crater (15 million years) with its ejecta blanket. The Rochechouart structure of similar size near Limoges, France, is 160 million years old, more than ten time the age of the Ries Crater. It can no longer be recognized as a crater in the field, but only by the occurrence of impact breccia and suevite with their evidence of shock metamorphism. The same holds true for the craters on the Baltic Shield of northern Europe. Most of these are now represented by lakes formed in the crater bottoms, the former crater rims and ejecta blankets being no longer present. The Siljan Ring structure in Sweden (50 km in diameter) is the largest European crater and shows only the excavated crater basement. In the middle is a central uplift of granite in which shatter cones are found.

Table 6. Probable impact craters with diameters greater than 20 km

Name	Diameter (km)	Age (Ma)
Araguainha Dome, Brazil	40	<250
Azuara, Spain	30	<130
Boltysh, Ukraine	25	100
Carswell, Canada	37	117
Charlevoix, Canada	46	360
Clearwater East, Canada	22	290
Clearwater West, Canada	32	290
El'gygytgyn, Russia	23	3.5
Gosses Bluff, Australia	22	142
Haughton, Canada	20	21.5
Kamensk, Russia	25	65
Kara, Russia	60	57
Logancha, Russia	20	50
Manicouagan, Canada	100	210
Manson, Iowa, USA	32	61
Mistastin, Canada	28	38
Popigai, Russia	100	39
Puchezh-Katunki, Russia	80	183
Ries, Germany	24	14.8
Rochechouart, France	23	165
Saint Martin, Canada	23	225
Siljan, Sweden	52	368
Slate Island, Canada	30	<350
Steen River, Canada	25	95
Strangways, Australia	24	<472
Sudbury, Canada	140	1850
Teague, Australia	28	1685
Ust-Kara, Russia	28	57
Vredefort, South Africa	140	1970

Source: Grieve and Robertson, Terrestrial Impact Structures, Geological Survey of Canada 1987, Map 1658A.

As we have seen, very large meteorites are heated to such high temperatures upon impact that they are completely vaporized. Therefore, no projectile has

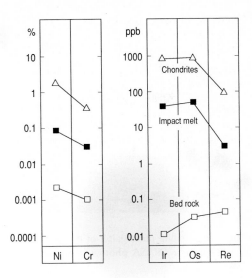

Fig. 31. Enrichments of meteoritic indicator elements in the impact melt from Clearwater-East. (After measurements by Palme, MPI Chemie, Mainz)

yet been found associated with the largest meteorite craters, and it is also futile to expect to find it at depth by drilling. The resulting vapor can, however, be condensed and deposited in the surroundings and become incorporated into the ejected and melted rocks. This provides a new possibility to prove the impact and even to determine the nature of the projectile.

1.6.5 Meteoritic Indicator Elements

The majority of meteorites have a chemical signature distinct from the rocks of the Earth's crust. They are,

Fig. 32. Satellite photograph of the impact craters Clearwater-West and -East, Canada. (NASA photograph)

for example, much richer in the noble metals iridium (Ir), osmium (Os), and rhenium (Re), and also in nickel (Ni). The incorporation of meteorite material, therefore, leads to a recognizable enrichment of these *meteoritic indicator elements* in the impact melt. In addition, elemental distribution patterns for various meteorite types can be distinguished, permitting, for instance, the determination of formation by a stony or iron meteorite. Figure 31 gives an example of the strong enrichment of the elements Ni, Cr (chromium), Ir, Os, and Re over the local rock for the Clearwater-East Crater (Fig. 32). The enrichment pattern corresponds to the distribution pattern of these elements in chondrites, indicating that the impacting body was a chondrite. Naturally, the meteorite distribution patterns are not always so easily interpreted. In the case of the Ries Crater an unambiguous determination of the impacting body has not been possible.

1.6.6 The Tunguska Event

Finally, a more detailed description will be given of an event that certainly represented the descent of a giant meteorite. The target area could be identified, but neither a crater not recoverable meteorite material was found. This occurrence, which fortunately took place in a completely uninhabited region of the Siberian taiga on the Stony Tunguska River, presents many puzzles. There is no doubt, according to the large number of observers, that a giant meteorite descended in this region. On the morning of 30 June 1908 at about 6 a.m., passengers on the Transsiberian Railroad in Kansk saw a meteor the size of the Sun travel through the sky from south to north. After it had disappeared beyond the northern horizon they heard the report of thunder followed by several more reports. The thunder was so loud that the locomotive engineer stopped the train because he believed that there had been an explosion on the train. The impact of the meteorite was so intense that seismographs at great distances, in Irkutsk, Tbilise, and Tashkent in the USSR, and in Jena in Germany, recorded tremors. On the recording charts of instruments measuring air pressure in England and Germany air pressure waves corresponding to the time of the explosions were observed years later.

The air pressure wave was so strong in the fall area itself that at the Vanovara trading station, approximately 65 km from the center, window panes shattered and doors were forced from their hinges. Likewise, the tents of the Tungus people were blown away and their reindeer dispersed. No one, however, appears to have been killed.

Not until 1927, as a result of the zeal and scientific enthusiasm of the Russian meteorite investigator Kulik, were the necessary resources for an expedition to the supposed impact site made available. Kulik found the area that we regard today as the impact site. In the center of an approximately circular area of destruction is a swamp region with several small lakes but without a meteorite crater (Fig. 33). Immediately surronding the swampy area is the *burned zone*. Here, one finds clear evidence of burning of the original stand of trees. This area extends about 20 km southeast of the swamp area. Surrounding this area is a *zone of flattened trees*. In this zone the trees were snapped off at the bottom as if they were reeds. The trunks lie aligned, pointing radially outward from the swamp area, as is shown by both ground observations and aerial photographs (Figs. 33 and 34). This zone stretches about 40 km to the southeast. For a part of the forest only the tree tops were snapped off. This regions appears as a gruesome, dead forest of telegraph poles (Fig. 35). Beyond the borders of this outer area, the forest gradually transforms into a normal stand of trees. The destructive pressure wave even reached the Vanovara trading post, as one finds occasional trees there with their tops snapped off.

Soil samples from the area of severe damage as well as from the surrounding 100 to 200 km produced upon investigation small magnetite and silicate spherules. They were probably produced by the fall of the meteorite, but no meteorite material as such has ben found. No impact crater exists. One assumes, therefore, that the penetrating body exploded above the ground at a great height. Perhaps it was the nucleus of a comet of relatively low density. There has also been speculation that it was caused by *antimatter.*

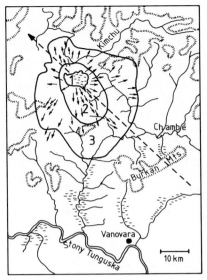

Fig. 33. Area of the Tunguska event with the three zones of destruction: *1* swamp with dead trees; *2* zone of forest fire: *3* zone of forest destruction. The *short arrows* shown the direction of fallen trees. The *long arrow* gives the direction of the fireball. (After Krinov, Giant Meteorites, Pergamon Press 1966)

Fig. 34. Devastated forest. (After Kulik, in: Nininger, Our Stone-Pelted Planet, Boston 1935)

Fig. 35. Telegraph pole forest. (After Kulik, Atlantis 1929)

In this (hypothetical) form of matter the nuclei of the atoms are negatively charged, surrounded by positive electrons. If antimatter and matter come together, both are annihilated in an explosion and nothing remains. There is absolutely no evidence, however, supporting this hypothesis.

1.6.7 Tektites

Modern meteorite impact research has brought another geologic puzzle nearer to solution: the tektite problem. Tektites are centimeter-sized, roundish glass bodies, mostly green or brown to black (Figs. 36–38), that were apparently melted (from the Greek *tektos* = melted). Their occurrences resemble those of meteorites, being found in strewn fields as individual specimens that have no relationship to their surroundings. The strewn fields are, to be sure, very large,

Fig. 37. Tektite from Billiton, Indonesia (slightly reduced)

Fig. 36. Moldavite (2/3 natural size)

Fig. 38. Button-shaped australite (natural size)

much larger than that produced by the largest meteorite shower. There are four known occurrences: the moldavites of Czechoslovakia, the tektites of the Ivory Coast, the gigantic strewn field from Australia, Indochina, and the Philippines, and the North American tektites from Georgia and Texas. For all of the land occurrences, with the exception of the moldavites, there are also associated microtektites from the neighboring oceans. These glass spherules (less than 1 mm in diameter) are found in specific layers of the sea sediments.

All the occurrences of tektites have similar chemical properties. They are highly siliceous (60 to 80% SiO_2), contain in additon about 10% aluminium oxide and a few percent iron, magnesium, calcium, sodium and potassium. They are therefore different from any

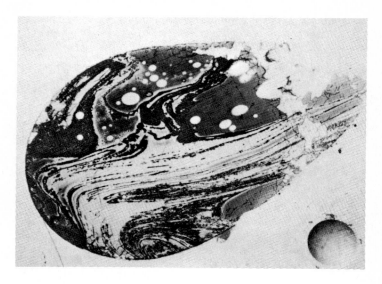

Fig. 39. Section through a glass drop from the Lonar meteorite crater, India. Length 0.7 mm. (After Fredriksson et al., The Moon 7, 1973)

known meteorites and also from terrestrial volcanic glasses. They are similar in major and trace elements to terrestrial sandstone, graywacke, or also loess, as was shown by E. Preuss as early as 1935. The similarities have since been shown to include the rare earth elements, rubidium and strontium isotopes, and lead isotopes.

L.J. Spencer put forth the theory in 1933 that tektites originated from the impact of a large body on the surface of the Earth. As a consequence, melt droplets would be sprayed over great distances producing a strewn field. He supported this idea with the melt glass that is found in association with many meteorite craters. Figure 39 shows a glass drop from the Lonar Crater in India that is similar to microtektite glass. Today, the theory of Spencer is generally accepted, and for at least two of the tektite strewn fields the craters of origin have been identified. The Ries Crater is the source of the moldavites found at a distance of 300 to 400 km from it, and the Bosumtwi Crater in Ghana is the source of the Ivory Coast tektites. This was proven primarily by W. Gentner and his coworkers in Heidelberg, who demostrated that the craters and their associated tektites had the same ages: 15 million years for the Ries Crater and the moldavites, and 1.2 million years for the Bosumtwi Crater and the Ivory Coast tektites. These ages were determined by physical methods that will be described later in Section 3.1. These results have made the alternate hypothesis highly improbable, namely, that tektites originated either by meteorite impact or volcanic eruption on the Moon and then falling to Earth. Now that we have been to the Moon and studied lunar rocks, it is even clearer that tektites do not come from the Moon.

Still unexplained is how tektites originate from an impact and how they reach their recovery place. Perhaps they do not originate directly as molten drops of glass, but condense from a cloud of vaporized material. This is the view of the mineralogist W. von Engelhardt, because the moldavites show enrichments in certain elements of large ionic radius such as calcium, strontium, and barium. This could be explained by the fact that large ions are preferentially incorporated in the silicon-oxide lattice during its formation by condensation.

1.6.8 Impacts and the History of the Planets

The inner planets of our Solar System were formed by the accretion of smaller bodies, from dust particle size to more than 1000 km in diameter. In the final phases of accretion the surface of the planet was shaped by these impacting bodies. Like the surface of our Moon, Mercury and the oldest regions of Mars are strewn with impact craters. Most of the lunar craters originated in the early history of the Moon 4.5 to 3 billion years ago. Lunar rocks have the same kind of shock-induced transformations seen in terrestrial impact craters, thus confirming the impact origin of these craters on Earth.

Meteorites have also been transformed on the surfaces of their parent bodies by other impacting bodies. In fact, for the chondrites, the most abundant representative of the stony meteorites, transformation through impact is the major "geological" process that has influenced their otherwise primitive structures. Examples are the formation of breccias from debris of various types, the formation of glass and melted rock, shock veins, and blackening (black chondrites). Perhaps chondrules also formed as impact-produced melt droplets that solidified as spheres. In any case, chondrule-like melt droplets in lunar soil and from the Ries Crater are known (Fig. 40).

The primitive Earth surface was certainly dotted with a similar concentration of impact craters as that of the Moon. Evidence of these craters was lost early through geological activity such as mountain building, erosion, and weathering. It is probable, however, that also in later Earth history impacts of asteroids and comets brought about catastrophic geological and

Fig. 40. Droplet chondrule from the suevite at Aufhausen, Ries crater, Germany; thin section, diameter 0.1 mm. (After Graup, Earth Planet. Sci. Lett. 55, 1981)

biological upheavals. A much discussed occurrence is that at the turn of the Cretaceous to the Tertiary period. The idea of an impact was born when in the thin clay layer at the boundary between these two ages a worldwide enrichment of the meteoritic indicator element iridium was found. It has been long known that 65 million years ago, at the time represented by this boundary, many types of animals became extinct, particularly the dinosaurs but also many species of marine plankton. An asteroid falling into the sea could have created a giant cloud of water vapor and dust that changed the atmosphere and the climate so severely that many species of animals and plants could not survive. Model calculations suggest than an impacting body 3 to 10 km in diameter would be necessary, creating a 150 to 200 km crater. A circular structure at Chicxulub in Yucatan, Mexico, is discussed as the possible impact crater. However, evidence that this structure was actually formed by impact is still rather scarce.

1.6.9 Earth-Approaching Asteroids

Whether the Cretaceous/Tertiary boundary really represents an impact catastrophe which caused the extinction of the dinosaurs is till controversial. It is certain, however, that the Earth has experienced impacts of this magnitude from time to time, and will again in the future, as asteroids are known to be in orbits that cross the orbit of the Earth. About 30 of these *Apollo* asteroids are known (diameters between 0.2 and 8 km); and their actual number is estimated to be between 750 and 1000. Although these orbits, which originate in the asteroid belt, extend within the Earth's orbit, an individual orbit only crosses the Earth's orbit when it lies in the same plane as the Earth. This is generally not the case. Their highly eccentric orbits change constantly under the influence of Jupiter, and calculations show that a typical Apollo orbit assumes a path that cuts the Earth's orbit once in every 5000 years. Of course, a collision only occurs when both bodies meet at the point of intersection of the orbits at the same time. Calculations indicate that a collision with a body of 500 m diameter (or larger) would happen once every 100000 years. Substantially more collisions are possible, however, with the small fragments of Apollo debris which results from the collisions of Apollo asteroids with one another. These fragments have the same orbit as their source body, but over time they gradually distribute themselves along the path.

Therefore, it is assumed that the majority of the meteorites that fall to Earth come from Apollo asteroids.

Small Earth-approaching asteroids represent a considerable threat to mankind. In order to know more about these bodies, a special *Spacewatch* telescope was installed at Kitt Peak in Arizona, USA. In 1991 it detected a body of 5–10 m diameter which passed the Earth at a distance of only 170000 km, half the Earth–Moon distance. If it had collided with the Earth, its explosive power would have amounted to about 400 kilotons of TNT, about three times the energy of the Hiroshima atomic bomb. Up to June 1993 40 more Earth-approaching asteroids have been observed with the Spacewatch telescope. Most spectacular was the discovery of the 5-km-diameter asteroid Toutatis in December 1992. It passed the Earth at a distance of 4 million km. It will return every 4 years, i.e., the next time in 1996. If it collides with the Earth, it would be capable of creating a biosphere disaster similar to that described from the K/T-boundary and would probably wipe out human civilization. Space missions to such objects could be launched in order to try to change their orbit so that they will not hit the Earth. Space missions to Earth-approaching asteroids, which are not directly on a collision course, could also study their composition not only for clues to their origin and evolution, but also because they may become a source of raw materials for space constructions in the future.

These very large meteorites represent the upper end of the spectrum of meteorites that fall to earth. The smaller meteorites are naturally much more numerous, and we will now come back to these "normal" meteorites. We will consider their distribution in terms of place and time.

1.7 Number and Distribution of Meteorite Falls

1.7.1 Frequency of Meteorite Falls

We have become acquainted with meteorite falls and are now prepared to ask further how many such falls are known to have produced material that has been recovered. Presently, material from about 950 observed falls has been recovered, but this is not the only material we have for study today. In addition to these 950 falls, there have been about 1800 meteorites found that were not observed to fall, thus totaling about 2750 known meteorites. Each year there are new falls and finds. Table 7 gives a summary of the representatives of the various meteorite types that have been identified to date.

We have already explained the difference between stone and iron meteorites (Sect. 1.3). The other classification designations given in Table 7 will be explained in Section 2.5. Table 7 shows that stony meteorites are far more numerous than the irons, and that among the stones the chondrites are the most abundant. There is also a preponderant type among the iron meteorites: the octahedrites. Conspicuous also is the difference between finds and falls: stony meteorites have about the same number of observed falls and finds, while 683 iron meteorites have been found compared to only 42 observed falls. The origin of this striking difference can be explained by the fact that stony meteorites are similar to some

Table 7. Number of meteorite falls and finds as of 1 January 1992

Class	Falls	Finds[a]	Total
Meteorites, total	952	1798	2750
All chondrites[b]	822	1023	1845
H-chondrites	293	485	778
L-chondrites	337	395	732
LL-chondrites	68	35	103
E-chondrites	13	10	23
C-chondrites	36	24	60
All achondrites[b]	70	25	95
Eucrites	23	9	32
Howardites	18	4	22
Diogenites	10	0	10
Aubrites	9	1	10
Ureilites	4	7	11
SNC meteorites	4	2	6
All stony iron meteorites[b]	11	61	72
Pallasites	4	37	41
Mesosiderites	6	23	29
All iron meteorites[b]	49	689	738
Octahedrites	28	457	485
Hexahedrites	5	44	49
Ataxites	0	33	33

Source: British Museum Catalogue of Meteorites (1985) and Meteoritical Bulletin 63–72 (1985–1992).

[a] The Antarctic finds are not included here. Finds from desert areas (Sahara, Nullarbor, Roosevelt County) were divided by two to account for possible pairings.

[b] The total numbers also include ungrouped members.

terrestrial rocks, and are also much more susceptible to weathering. Iron masses, on the other hand, generally resist weathering. They are such obvious foreign bodies on the Earth's surface that even when found long after their fall they are comparatively easily recognized as meteorites.

1.7.2 Geographic Distribution

The nest question to ask is, do meteorites fall preferentially in certain regions of the Earth's surface? On the map in Fig. 41 the distribution of recent meteorite falls (listed in Sect. 4.5) is shown and, in addition, the total number of observed falls per km^2 for different continents and countries (up to 1992). This distribution does not have its origin in the preferential fall of meteorites in certain regions, but in the density and cultural circumstances of the population. The highest fall numbers are observed in the most densely settled areas: Europe, India, Japan, whereas less populated countries like Australia show the lowest number of falls per km^2. The low numbers in Africa, South America and China are also related to the lower levels of education and communication in these countries. The high number of recent falls in China (Fig. 41) shows that this country is now catching up.

How many meteorites really fall to Earth? Their number can be estimated from the photographic records of meteor trails. In Canada between 1974 and 1983 meteor trails were recorded over an area of 1.26 million km^2. Forty-three trails that must have produced meteorite falls were observed (only one was recovered, see p. 12). From the velocity, the brightness, and the endpoint of the trail, the mass can be calculated; the values given in Table 8 were thus obtained.

In reality, however, only one to two dozen meteorites worldwide are observed or found each year, meaning that by far the great majority remain unobserved. This is true even for heavily populated areas such as Europe. In its surface area of 10.1 million km^2, about 90 meorites of more than 1 kg must

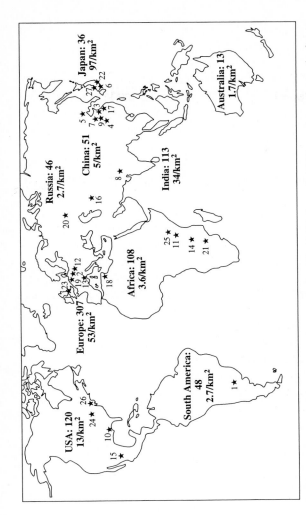

Fig. 41. World map: *stars* show recent meteorite falls (1985 through 1993) *numbers* refer to the list in Appendix 4.5. Also given are the total number of observed meteorite falls (up to 1993) for different continents and countries, and the falls per km²

fall yearly. In the 8 years from 1985 to the end of 1992 only five falls were observed, that is less than one fall per year (see Sect. 4.5).

This yield, however, can be improved, as was demonstrated by H.H. Nininger in the USA. He was stimulated by the fireball of 1923 in the Midwest of the USA to search for the fallen meteorite. He did not find that meteorite but did recover a selection of others. This led him to give up his teaching position in 1930 and devote himself completely to collecting meteorites. Through lectures and articles he educated rural people about meteorites and their recognition. By working in the prairie states where the flat, rock-poor landscape was favorable for finding meteorites, he succeeded in time in recovering a large variety of meteorites including some new falls. Between 1923 and 1949 he found 37 new meteorites in Kansas, a region in which previously only 15 were known. Wyoming, Texas, Colorado, and Nebraska provided similar results. Nininger was especially active in the investigation of Meteor Crater and in the recovery of the associated Canyon Diablo meteorite material from its environs (Fig. 42).

Table 8. Number of meteorite falls per year, calculated from fireball observations

| Area | Weight per fall | | |
	> 0.1 kg	> 1 kg	> 10 kg
1 million km^2	58	9	1.3
Land surface of the Earth	8600	1300	200
Total Earth surface	28000	4400	700

Source: Halliday, Blackwell and Griffin: Meteoritics 24 (1989).

Fig. 42. Harvey Nininger and his son Bob at Meteor Crater, Arizona. With this self-built magnetic rake they collected in 1939 about 12 000 small iron meteorite fragments, total weight 20 kg, from a 2-mile zone around the crater. (From Nininger, Arizona's Meteorite Crater, American Meteorite Museum, 1956)

In the past years a completely new source of numerous meteorites has been discovered: Antarctica. After sporadic, individual finds over the years, a Japanese field party in the Yamato Mountains in 1969 found nine meteorites on bare ice within an area of 50 km². Subsequent work in the same small area produced approximately 1000 meteorites in 1974/1975 and an additional 3000 in 1979/1980. American expeditions that have included international participation have been mounted yearly starting in 1976 from the US Antarctic base in Victoria Land. By the middle of 1987 a total of 8900 meteorites had been found. The most abundant finds are chondrites (95%). There were also 158

Fig. 43. Meteorites on a blue ice field in Antarctica, found during the 1983/1984 field season at the Allan Hills. (Photograph L. Schultz, Mainz)

achondrites found that have contributed importantly to the study of these rare meteorite types.

The Antarctic meteorites are found on so-called blue ice fields (Fig. 43). These are snow-free regions where ice is forced up by an underground barrier such as a mountain ridge, and where it is ablated by the constant wind. Meteorites that have fallen over a larger area are carried along within the ice and become concentrated when they emerge on blue ice surfaces (Fig. 44). The destruction of meteorites is much

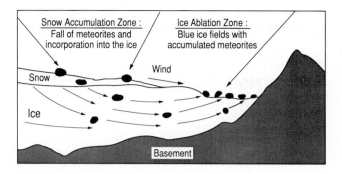

Fig. 44. Model for the meteorite concentration mechanism in the Antarctic, (After L. Schultz)

slower here than in other regions of the Earth. Measurements have demonstrated that most of these meteorites have been in the ice up to several 100000 years. The Antarctic meteorites thus provide a sampling of the meteoritic material of our Planetary System over a much longer period than the non-Antarctic falls that have come to us only in the last 200 years. Older observed falls have not been preserved and meteorites in the ground disintegrate over time. The determination of the terrestrial ages of these meteorites (see Sect. 3.1) is also important for ice research in the Antarctic, as they provide an approach to determine the age of the ice. Thus, it may be possible to find explanations to its formation and motion in the past and perhaps even to explain climatic variations.

Meteorites that definitely originated from the Moon were found for the first time in Antarctica. We will return to these in Section 3.2.2.

Statistically, meteorite falls should be distributed equally over the Earth with a slight decrease in polar regions. Their spatial distribution is thus random.

Fig. 45. Staff of the Natural History Museum, Washington, USA, examining the second meteorite that fell at Weathersfield in 1982. From *left to right*: Dr. Brian Mason, Dr. Roy Clarke, Tim Rose, and Twyla Thomas. (Photograph Smithsonian Institution)

Nevertheless, this can result in highly improbable occurrences, for example, in the small city of Wethersfield, Connecticut (USA). There on 8 April 1971 a meteorite fell through the roof of a house and 11 years later on 8 November 1982 a second fell, which also penetrated the roof of a house (Fig. 45). Also noteworthy is that three of the largest iron meteorites ever found (see Fig. 14) are located in the vicinity of the fall ellipse of the large meteorite shower of Allende, northern Mexico. Here, as in all meteorite finds, climate and vegetation play a decisive role: e.g., in dry areas with little vegetation meteorites deteriorate more slowly and can be found more easily.

1.7.3 Times of Fall

The distribution of meteorites over time will now be considered. Can a pattern for the month of fall be established? The answer to this question is significant because it could lead to inferences concerning relationships to meteor showers or comets. Figure 46 presents the distribution by month of well-documented, observed meteorite falls for our northern hemisphere through the end of 1986. One sees immediately that falls of all meteorite types were observed in every month. This means that there are no "meteorite streams" through which the Earth passes in a given month as with shooting stars and meteors. There is, however, a clear maximum in the months of April, May, and June. Is this a real effect in that more meteorites actually fall, or is it due to the fact that more people are out of doors in these early summer months and the probability is greater that a fall will be observed? A better picture of the actual fall rate distribution can be obtained from the photographed meteorite trails that were determined by the three previously mentioned camera networks. Trails of 74 meteorites were selected whose characteristics suggest that they must have resulted in falls of meteorites. Their distribution over the year is plotted as an additional curve in Fig. 46, normalized to the maximum of observed falls in May. The shape of the curve then shows how many falling meteorites can be expected to be observed in the other months if the relationship remains constant. We see that the calculated curve has a maximum in spring and that it runs parallel to the observed falls until September. During the winter months up to March, the curve for fall observations is substantially lower. This shows that this reduction must be due to the poor observing conditions during winter. The maximum frequency of falls in spring, calculated from meteorite trails, is a more important finding, as meteor showers and comet residues are at a minimum during this period and reach a maximum in July through September. This means that meteor showers and meteorite falls are probably unrelated.

The relationship of meteorite falls to the hour of the day is also of interest (Fig. 47). The first thing noticed is that substantially more meteorite falls are observed during the daylight hours from 6 to 18 when most people are awake than during the night hours. If, however, we compare the daylight hours from 06:00 to 12:00 h with those from 12:00 to 18:00 h, during which the visual conditions should be the same, there are actually more falls in the afternoon (301) than in the morning (157). This means for most meteorite types there are always about twice as many falls observed in the afternoon as in the morning: for the H-chondrites it is 92 falls to 48, for the L-chondrites 117 to 49, and for all chondrites together 250 to 119. Interestingly, this does not apply to the achondrites. For this group the numbers are approximately equal, 20 falls to 19.

These times of fall are important since they give clues to the direction of motion of the meteorite. All those that fall from midday to midnight move in the same direction as the Earth, while those that fall from midnight to midday come from the opposite direction or are overtaken by the Earth. Figure 48 demonstrates this situation. This is naturally of importance for the velocity with which the meteorite enters the Earth's atmosphere. When moving in the same direction, the

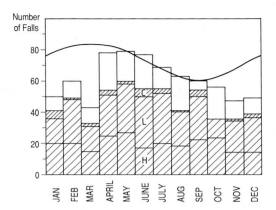

Fig. 46. The monthly distribution of 731 observed meteorite falls in the Northern Hemisphere (statistics by B. Klein after data from the BM Catalogue of Meteorites). The *hatched area* gives the number of classified chondrites, subdivided in the H-, L-, and C-groups. The curve was calculated by Halliday and Griffin (Meteoritics 17, 1982) from recorded meteorite trails

Fig. 47. The hourly distribution of 671 observed meteorite falls (statistics by B. Klein after data from the BM catalogue of Meteorites). The curve was calculated by Halliday and Griffin (see Fig. 46), it was normalized to the maximum of the observed falls between 14:00 and 18:00 h

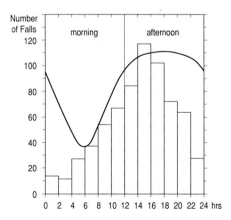

entrance velocity is the difference between the velocity of the Earth (average of 29.77 km/s) and that of the meteorite. Approaching the Earth from the opposite direction leads naturally to the addition of the two velocities. The larger the velocity upon entering the atmosphere, the greater the amount of heating and melting the meteorite experiences.

In Fig. 47 we again have the distribution of observed falls superimposed by the calculated curve from the photographed meteorite trails. This has a pronounced minimum at 06:00 h, the time which corresponds to the Earth's apex (front side) on its path around the Sun, and a broad maximum between 15:00 and 21:00 h on its opposite side. This shows that the majority of meteorites circle the Sun in the same direction as the Earth; they hit the Earth from behind while overtaking it. We see that the increase in falls from 06:00 to 15:00 h runs parallel to the calculated curve, but a strong deficit is seen in the evening and night hours.

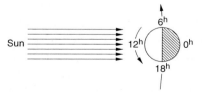

Fig. 48. Local times on the Earth in relation to its rotation and its orbit around the Sun

1.7.4 The Dangers of Descending Meteorites

A practical aspect of meteorite falls, which directly concerns individuals, will now be considered. We have already seen that the damage caused by an ordinary meteorite is insubstantial, despite the thunder and lightning accompanying the event. Nevertheless, when one realizes that many meteorites weigh over 1 kg and meteorite showers sometimes produce up to 100000 individual stones, one can imagine that it is not the most pleasant situation when one has to be present at the site of a fall. History and statistics, however, offer comfort in this case. To date, there is no record of a single, individual, authenticated case in which a person has been killed by a falling meteorite. To be sure, there are stories told, where this is claimed: in 1511 in Cremona, Italy, and in 1650 in Milan, a monk was supposedly killed by a meteorite, and in 1674 two Swedish sailors were supposed to have been struck on their ship. All of these older reports as well as those of more recent times – a few years ago in a Balkan country a wedding guest in a wagon, or a child in Japan, or finally in 1906 the rebellion leader, General Catillianis, in his camp were reported to have been killed by meteorites – have not withstood critical investigation. Perhaps the most endangered were three children in Braunau in Bohemia. In 1847 a 17-kg iron meteorite fell into the room in which they slept and the rubble from the roof covered them, but without causing any serious injury. A woman was slightly injured by the fall of the Sylacauga, Alabama, meteorite of 30 November 1954. The 3.8-kg stone penetrated the roof of ther house, the ceiling of the room, then bounced off a radio, subsequently injuring her leg.

Of course, it is certainly possible that at some time a person will be killed by a meteorite. The probability, however, is very small. It must be recalled, that people's bodies represent an exceedingly small surface area when compared to the total surface area of the Earth. We can thus calculate from the figures in Table 8 that in the USA the probability of a person being hit by a meteorite of more than 1 kg is once in 1000 years. A comforting estimate! It should also be noted that large showers have fallen in heavily settled regions without anyone being injured. An example is the meteorite shower which hit the small town of Mbale, Uganda, in August 1992. More than 50 stones of up to 10 kg fell over an area of a few km^2, but nobody was injured. A boy was hit on the head by a little stone of only 4 g, but he was not hurt. Buildings, on the other hand, are more commonly damaged, for example, by the meteorites from Ausson (1858), Barbotan (1790), both in France, Benares (India, 1798), Braunau (Bohemia, 1847), Mässing (Bavaria, 1803), and Pillistfer (Estonia, 1868). It seems that a high proportion of the recent meteorite falls has damaged buildings: of the last ten meteorites that fell in the USA prior to 1986, one (Louisville) hit three buildings and a car, five others hit buildings, and two hit other structures. However, these meteorites were recognized *because* they damaged buildings or other structures; many other falls, which did not, remained unnoticed. An example is also the meteorite in Kiel, Germany, in 1962 which fell through a roof (Fig. 49) and was found the next day in the attic of the house, no other signs of its fall had been noticed.

We have only a few reports of animals that were killed by meteorites. For example, a dog is supposed to have been killed by the Nakhla, Egypt, meteorite fall

Fig. 49. Tin roof with hole (diameter about 10 cm) made by the impact of the Kiel meteorite, weight 738 g, on April 26, 1962. (After Schreyer, Natur und Museum 94, 1964)

of 1911 and a foal by a stone from the New Concord, Ohio, meteorite shower of 1860.

1.8 The History of Meteoritics

Spectacular displays accompanying the fall of a meteorite not only produce awe and fear in people today, but this type of response was naturally much more intense in the past when our view of nature and the state of the heavens was based on religion, myth, and Aristotelian concepts. Each brightly illuminating and rapidly traveling star produced the impression that it fell out of the eternally silent crowd of its brothers. These events aroused great interest, especially when some of these stones that had fallen to Earth were found and collected. The history of the recognition of the true nature of meteorites is of particular interest, a warning example for scientists, reminding them to approach natural phenomena critically and without preconceived notions.

The scientific study of meteorites is only about 200 years old, but the knowledge of meteorite specimens goes much further back in time. Meteorites were probably used as a source of metal to manufacture objects of various types long before man was able to smelt iron ores, as it was the case not long ago with Eskimos and various African tribes. Drawing on results from the comparative study of languages, we know that the word *iron* in many languages is related to the concept of *heaven* or *star*, for example, the Greek word for iron (*sideros*) and the Latin word for star (*sidera*). In Egyptian the word for iron means *metal from heaven*, and in the inventory of the treasure house of a Hittite king iron is recorded as *iron from heaven*, while gold and silver are listed from various mines. Other native peoples, the Australian aborigines and most of the American Indian groups, for instance, do not seem to have made use of these heavenly gifts, although meteorites are found fairly frequently in the lands they inhabit. Ancient veneration and utilization of meteorites in the Old World explain why fewer iron meteorites are now found there than in the New World. The prehistoric iron meteorite shower that fell on the Island of Ösel, which has been inhabited for a long time, has only produced paltry remains after intensive searching and digging. On the other hand, in Namibia (southwest Africa), Mexico, and Australia tons of iron meteorites were available without their usefulness being appreciated.

The ancient civilizations, Chinese, Egyptian, Greek, and Roman, had a lively interest in meteorites and have left behind many records concerning their falls and finds. Mu Tuan Lin (1245–1325) collected approximately two centuries of meteorite fall reports in the great Chinese encyclopedia. In Nogata, Japan, a meteorite has been preserved as a treasure in a Shinto shrine for over 1000 years. The old records report that it fell on temple grounds, accompanied by lightning and thunder, on 19 May 861. An investigation undertaken in 1982 established that it is a meteorite, an L-chondrite. It is, therefore, the oldest observed meteorite fall from which material has been preserved. For many years this honor belonged to the Ensisheim fall of 1492 in Alsace.

In the writings of Anaxagoras, Plutarch, Livius, Pliny, and other ancient writers reports on meteorite falls are found. Only a few examples will be given here. In approximately 625 B.C. a rain of stones fell in the Albani Mountains near Rome, in 467 B.C. a large stone fell in Thrace at Aegospotami. In about 204 B.C. a stone that had fallen earlier in Phrygia was taken to Rome with great ceremony. The meteorites were viewed as stars which had fallen to Earth or as messengers from the gods, and were for this reason often treated as objects of religious veneration. The statue of the goddess Diana in Ephesus and the sacred relic in the Temple of Venus in Cyprus were said to have been meteorites. In Rome at the time of Numa Pompilius an iron meteorite that had fallen from heaven was venerated in the form of a small shield. It was believed that its possession was

83

connected with world dominion. In the desire to discourage thievery, the clever Roman priests manufactured 11 more iron shields having the exact form. Occasionally, meteorite falls were perpetuated on coins. Figure 50 reproduces such a coin that commemorates the death of Caesar. The Kaaba stone in Mecca, a dark object mounted in silver, has been claimed by some to be a meteorite (recent investigations, however, have found no indication that this is true). On occasion, American Indians have given meteorites ritualistic treatment. This explains the burial of the 24-kg Winona stone meteorite in a stone box found in the ruins of the Elden Pueblo in Arizona.

Weapons of all types have been made out of iron meteorites since ancient times. Sword blades from Arabia are known that were said to have granted their owners invulnerability. The Mogul ruler Dsehangir in 1621 had sabres, daggers, and knives manufactured from meteorites. More recently, the Sultan of Solo in Java had kris (Malay daggers) produced from the Prambanan iron meteorite (known since 1797) and used them as princely gifts. Meteorites were much used in the manufacture of knives and nails. Large meteorite blocks served as anvils, examples are the two

Fig. 50. Meteorite coin, ancient Rome

Fig. 51. Woodcut from 1492 showing the meteorite fall at Ensisheim, Alsace

Tucson masses in the Smithsonian collection in Washington. Many of the iron meteorites in our collections show evidence of heating and working at the forge, abusive treatment from a scientific point of view as opposed to their cultural value.

During the Middle Ages, men viewed meteorites, and similarly comets, as messengers from God, but in contrast to antiquity, as indicators of God's wrath. The large Elbogen iron meteorite from Bohemia was known during the Middle Ages as "the bewitched baron". According to the legend, the bad baron of the castle of Elbogen near Carlsbad was transformed into this iron (weight of 107 kg). The meteorite was first preserved in the dungeon, and later in the Elbogen town hall, where a piece of it is still exhibited.

The oldest European meteorite fall from which material is still preserved is the fall at Ensisheim in Alsace. At 11:30 a.m. on 16 November 1492 after a violent detonation, a stony meteorite fell to the

85

Fig. 52. The Ensisheim stony meteorite on display in the town hall

ground. Figure 51 reproduces a broadsheet of the time describing the fall. The German poet, Sebastian Brant (1457–1521), composed a Latin poem commemorating the event. He writes that the fireball was seen over Switzerland, Burgundy, and the upper Rhine Valley, boding ill for the whole area. The largest part of the stone is still preserved today in the town hall in Ensisheim (55.75 kg, see Fig. 52). Kaiser Maximilian used the fall as a sign from God in an exhortation against the Turks. The fall of a stone meteorite in the Ortenau basin SW Germany, in 1671 was considered as an expression of rage from God against " the cold hearts of the Turks and fierce acts they committed against true Christians".

The learned people of the time did not know what to do with the stone from Ensisheim. They concluded that the fall was simply one of God's miracles. In later times, particularly during the Enlightenment (18th century), the scientific world rejected these beliefs in the miraculous. And since they found in the old reports of meteorite falls much that was false, exaggerated, and fanciful, they also banished without hesitation the well-authenticated material to the realm of fairy tale and absurdity. The French Academy was particularly influential. The meteorite which fell in Lucé, France, in 1768 was investigated by one of its commissions and declared to be pyrite, singed by lightning. With regard to the well-observed stone meteorite shower of Barbotan (France, 1790), reported by the mayor and city council, the French scientist Bertholon wrote: "How sad it is that a whole municipality through an official report containing all manner of folk-tales certifies what not only physicists but all sensible people see as pitiable". And in Germany, X. Stütz wrote: "Indeed, in both cases [the meteorites from Hraschina, Croatia, 1751, and Eichstädt, Bavaria (Germany) in 1785] where iron is supposed to have fallen from the heavens, it was still possible in 1751 in Germany that this was believed, even by enlightened individuals, due to the prevailing ignorance concerning natural history and experimental physics, but in our time it would be inexcusable to consider such fairy tales as even probable". In spite of this refusal of acceptance, Stütz considered these two meteorites so remarkable that he saved them along with others and thereby founded the basis of the famous Vienna Meteorite Collection.

The physicist, Ernst Florens Friedrich Chladni (born in 1756 in Wittenberg, died in Breslau in 1827), who is well known for his contributions to acoustics, first had the courage to challenge the prevailing authoritative view. Among his activities on his many trips he collected reports on meteorite falls, and in his small book that appeared in 1794, "About the origin of the iron mass found by Pallas and other similar iron masses and about some associated natural phenomena", he brought unfamiliar data together with his own observations on meteorites and concluded that they came from fireballs and must originate beyond the Earth. His view received immediate resistance and was much ridiculed. In fact, the beliefs of the time were so grounded in mysticism that even the physicists, whom he felt might believe him, made fun of him instead. His famous colleague Lichtenberg expressed his opinion of the book: "When I read this text, I felt as if I myself had been hit on the head by such a stone". Others judged Chladni as "one of those who deny the orderly nature of the world and do not consider how guilty they become of all the evil in the moral world". Nature, however, came to Chladni's aid. A large shower of meteoritic stones that was observed by unimpeachable witnesses fell in L'Aigle (France) in 1803, compelling its recognition by the Academy in Paris.

It has become common practice to condemn the narrow-mindedness of the natural scientists before the acceptance of Chladni's views because of their bias in accepting meteorite falls. One should be cautious, however. None of these scientists had personally witnessed a meteorite fall. Also, Chladni himself could not present new and irrefutable observations. It was thus important to decide how far one could rely on reports presented exclusively by laymen. As pointed out above, these laymen accounts were often fanciful, uncritical, and false. Every modern meteorite investigator can recite many such experiences. This was the time that natural science was just beginning to free itself of all types of inhibiting influences. Its practitioners lacked knowledge of the developments of nearly 200 years that are available to meteorite students of today. They wanted to consider only what they themselves could confirm through observation. For them, it was obviously very difficult to distinguish the few correct reports from the overwhelming number of false ones. The problem of recognizing meteorite falls was, therefore, not one of natural science but a psychological one: it depended upon the correct evaluation of eyewitness reports. F.A. Paneth was the first to comment on this circumstance and noted that before Chaldni became a natural scientist he had been a jurist. He had been trained in a field that was deeply concerned with evaluating eyewitness accounts.

After Chladni's ideas were fully accepted, a lively interest in the collection and study of meteorites developed. They were soon of great interest, and at that time they were the only specimens available which provided data on the composition and properties of extraterrestrial bodies.

In the next chapters the compositional and structural nature of meteorites will be elucidated, now that we have become acquainted with the most important phenomena that occur when these heavenly bodies hit our Earth. First, however, we will briefly discuss two practical background topics.

1.9 Observations of Importance for a Meteorite Fall

Meteorite falls occur so seldomly and so suddenly that scientists have few opportunities to make observations.

We are therefore very dependent on the help of the general public. The suddenness of an occurrence results in untrained observers being completely surprised, and most of them experience such a fall only once in a lifetime. It is completely understandable, therefore, that the ensuing observational accounts are seriously deficient. Inconsequential details are considered as important while important observations are overlooked, forgotten, or inaccurately presented.

First, a general rule: all observations, particularly numerical ones, must be written down at the earliest possible moment. Few of us understand how unreliable memory is. Listed here are important points that deserve attention.

1. At what time (day, hour, minute) did the meteorite (fireball) appear? As soon as possible, compare the timepiece used to a source of accurate time.

2. How long did the light display last from its first sighting until it extinguished? Count the seconds (one thousand and one, etc.) as there is insufficient time to look at a watch. Counting as suggested is better than other estimates.

3. How bright and how large was the fireball? Compare with stars or the Moon.

4. What were the characteristics of the emitted light (sparking, flashing, exploding)? Make a sketch.

5. What were the colors of the fireball in the various parts of its path ?

6. Was a luminous trail, a smoke cloud, or any other track noticed ?

7. How long did the trail remain in the sky, its form and color ?

8. Where was the path of the fireball in the sky? An accurate answer to this question is of great interest. In starlit sky, use stars and constellations for orientation. In cloudy weather or during the day, use objects on the ground such as houses, church towers, trees, mountains, etc. Note exactly the point from which your observations were made. Height estimates in degrees are only useful when made by a practiced observer. Of greatest interest is the accurate determination of the endpoint of the trail.

9. How many seconds, or perhaps even minutes, were between the first appearance of the light and the arrival of the first sounds?

10. What was the nature of the sound (thunder, explosion, crashing, roaring, whistling, etc.)? How long did it last?

11. How much time elapsed between the fall of the meteorite and the first recovery of a meteorite specimen?

12. Was the meteorite hot or cold? Traces of burning? Smoke? An odor?

13. Did a single meteorite fall, or was there more than one? Record specimen weight(s) and dimensions. Make sketch(es) with dimensions of the recovery location from different directions, and note specimen weight and shape for each site. Make inquiries in the neighborhood. Individuals pieces are often found 1 km or further apart.

14. What was the nature of the ground and surroundings of the fall site(s)? Cultivated farmland, pasture, forest, road, sand — wet, dry, frozen, etc. Were trees or buildings damaged?

15. How deep did the meteorite penetrate into the soil? For several pieces give size, weight, and depth of each. Did the specimen break on landing?

16. What was the shape, dimensions, and inclination of the penetration channel?

17. Did the meteorite reach the ground before or after the accompanying sounds?

When photographs are taken, if at all possible, they should be taken from several angles and include something for scale (a person, a walking-stick, a hat, a coin, etc.).

1.10 How Can a Meteorite Be Recognized ?

There is no single characteristic that allows one to distinguish a meteorite from terrestrial rocks or artificial products under all circumstances. It is only possible to make this distinction by using a combination of several characteristics. These individual characteristics will be described more fully in Chapter 2. For stony meteorites they are:

1. The fusion crust (p. 111),
2. The characteristic depressions in the surface that are referred to as thumb prints (p. 110),
3. The presence of metallic iron particles enclosed in stone (p. 120),
4. The occurrence of small spherical bodies, the chondrules (Fig. 53 and p. 113).

Not all stony meteorites, however, have all of these characteristics. For instance, the achondritic meteorites have neither chondrules nor metal particles. On the other hand, there are terrestrial rocks that

Fig. 53. Broken surface of the chondrite Bjurböle (length 2 cm). Two large chondrules are visible (*left* and *upper middle*) as well as indentations from missing chondrules

contain metal. These are very rare, to be sure, with only two occurrences known: the basalts from Bühl near Kassel, Germany, and from Ovifak, Greenland. More numerous are terrestrial rocks with spherical structures, for example, oolite. These spherical bodies are formed differently than chondrules, however, and can be distinguished from them by their structure.

Iron meteorites may be recognized by the characteristic Widmanstätten pattern that appears on polished surfaces after etching with acid (Fig. 54; see also Sect. 2.5.2). This structure has never been observed in terrestrial or manufactured iron. There are, however, genuine meteorites known that do not display this pattern. Characteristic for all iron meteorites is the presence of both nickel (5 to 20%) and cobalt. Of course, naturally occurring iron may also contain nickel, but either too little (about 3% or less) or too much (more than 35%). The presence of nickel can

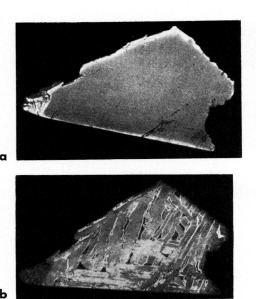

Fig. 54a, b. Iron meteorite Toluca, Mexico (length 3 cm). **a)** Polished; **b)** etched

be established by comparatively simple means, and a method for doing this is included in the Appendix.

If the reader feels that on the basis of the characteristics discussed here that a stone or a piece of metal is a meteorite, he or she should turn to a natural history museum in the area. Or better still, send the piece for a free examination to one of the institutes listed in the Appendix that employ specialists in the study of meteorites. One should not, however, be too hopeful. Experience has shown that for every 100 suspected meteorites received, at least 99 are something else. That 100th specimen, however, may be a specimen of great importance to science.

Therefore, we implore all those who have or will find a meteorite in the future to handle it with the care and respect that such a unique and irreplaceable object deserves. Every break, hammer stroke, or even the mildest of heating, greatly diminishes its scientific (and commercial) value.

2 The Meteorites

2.1 Cosmic Dust

Meteorites range extraordinarily in size and weight. At the low end of the scale is cosmic dust, a term that designates particles smaller than 0.1 mm. These small particles are not melted by frictional heating during passage through the atmosphere. Due to their large surface to mass ratio, the generated heat is lost by radiation. There are also melted particles among these micrometeorites; they formed mainly by ablation from the molten surface of larger meteorites. The Earth accumulates about 10000 t of cosmic dust per year, but the problem for meteorite researchers is to distinguish it from the tremendous quantities of terrestrial dust. Two approaches to this problem have been found:

1. The capture of cosmic dust at high altitudes in the atmosphere;
2. The recovery of cosmic spherules from deep-sea sediments, or from Arctic or Antarctic ice.

The first method employs American U2 aircraft that have successfully retrieved fluffy, porous particles from the stratosphere at altitudes of 20 km (Fig. 55).

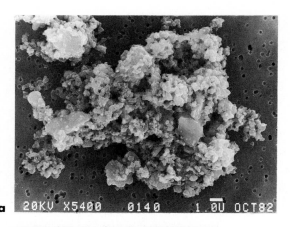

a 20KV X5400 0140 1.0U OCT82

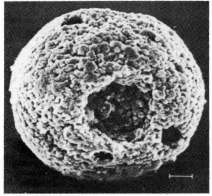

b

Fig. 55a, b. Cosmic dust particles, pictures taken with the scanning electron microscope. **a** Porous aggregate collected in the stratosphere, *scale bar* = 1 μm (after Bradley et al., Nature 301, 1983). **b** Deep-sea spherule, *scale bar* = 10 μm. (after Parkin et al., Nature 266, 1977)

They contain olivine and pyroxene, magnetite, water-containing silicates, and iron-nickel sulfide, minerals that are also found in the carbonaceous chondrites. Like these, they contain a few percent carbon and also have a similar bulk composition. They are, however, more fine-grained, a 10-µ aggregate containing up to 10^6 different individual grains. They are assumed to originate mainly from comets.

The deep-sea spherules (Fig. 55) can most easily be recovered from the sediments with a magnet. They are magnetic due to their magnetite content, and in addition contain mainly olivine with an occasional grain of metallic nickel-iron. For the most part, they originate from the surface melt of meteorites in the atmosphere, and they have bulk compositions similar to the melt crust of meteorites. Among these are a few that probably fall directly as micrometeorites.

Recently, concentrations of cosmic spherules have been found in the ice of Greenland and Antarctica. The spherules that fell into the very clean ice of interior Greenland were locally concentrated in shallow melt water lakes that formed in depressions. A French group has recovered cosmic dust by melting large quantities of ice in Antarctica. In $1\,m^3$ of ice an average of about 50 cosmic particles were found.

In any case, it appears that cosmic dust presents us with material that is not otherwise represented among the meteorites. Most important is the fluffy, porous material that could not survive passage through the atmosphere in large masses, for example, the material from comets. Perhaps it also contains interstellar particles from beyond our Solar System. Plans are underway to collect cosmic dust with large recovery instruments from space stations. It is expected that with the use of special detectors it will be possible to determine orbits and velocities of single particles so that conclusions regarding their origin can be drawn (Fig. 56).

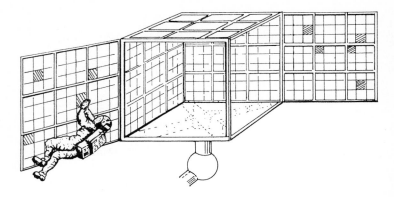

Fig. 56. Design for a cosmic dust collector to orbit around the Earth with a space station. On each side (3 × 3 m) of the cube nine different dust detection or collection experiments can be installed. (After McDonnell, LPI Tech. Rep. 86–05, Houston 1986)

2.2 The Size of Meteorites

Individual pieces of fist to head size predominate among the stony meteorites, with only an occasional specimen weighing more than 50 kg. A few meteorite showers have produced an abundance of the very smallest individuals, like the previously mentioned "Pultusk peas" from the fall near Pultusk in 1868. Iron meteorites are ordinarily much more substantial, with weights of 50 to 100 kg being common. The iron meteorites Treysa, shown in Fig. 57 (fall of 3 April 1916) weighed 63 kg with a diameter of 36 cm, and the iron North Chile (Buey Muerto; cf. Fig. 62), weighed 75 kg with a diameter of 40 cm. The largest meteorite fall that has ever been observed was the 23-t Sikhote-Alin iron meteorite shower in Siberia in 1947;

Fig. 57. Iron meteorite Treysa, Germany. About 1/8 natural size. (After Richartz, Schriften zur Beförderung ges. Naturwiss., Marburg 1917)

its largest member was 1.75 t. Table 9 lists some of the heaviest meteorites.

Illustrations and descriptive material follow to supplement Table 9. The heaviest stony meteorite known to date is the 1.8-t Jilin. It fell as one individual of a meteorite shower that produced 4 t altogether; it was recovered from the head (apex) of the ellipse of fall, in a hole 6 m deep. The Hoba meteorite in Namibia, weighing approximately 60 t, is the heaviest known iron meteorite and still lies where it was found (Fig. 58). Its nickel content is 16.2%, meaning that here there is a nickel reserve of 9.7 t and a cobalt reserve of 456 kg. Enterprising individuals in the past have attempted to recover this valuable supply of metal. Fortunately, the government of Namibia has declared this meteorite a natural monument to protect it from further damage.

The large iron meteorites from Cape York in Greenland have long been known by the Eskimos who lived in the region. Surrounding the 3-t block, "The Woman", a 1-m-high pile of basalt hammer stones has been found. With great difficulty generations of Eskimos have knocked off small pieces of metal that they fashioned with walrus bone into knives and harpoons. In 1894 they showed the North Pole explorer Peary

Table 9. The largest meteorites

Name	Date of fall or year of find	Class	Weight	Present location

1. Stony meteorites:

Jilin, China	1976/3/8	H chondrite	1.8 t	Jilin
Norton County, USA	1948/2/18	Aubrite	1.07 t	Albuquerque
Long Island, USA	1891	L chondrite	564 kg	Chicago
Paragould, USA	1930/2/17	LL chondrite	408 kg	Chicago
Bjurbole, Finland	1899/3/12	L chondrite	330 kg	Helsinki
Hugoton, USA	1927	H chondrite	325 kg	Tempe
Knyahinya, Ukraine	1866/6/9	L chondrite	293 kg	Vienna

2. Stony iron meteorites:

Huckitta, Australia	1937	Pallasite	1.4 t	Adelaide
Krasnojarsk, Russia	1749	Pallasite	700 kg	Moscow
Brenham, USA	1882--1947	Pallasite	450 kg, 215 kg 212 kg, 173 kg 162 kg, 100 kg	

3. Iron meteorites:

Hoba, Namibia	1920	Ataxite	ca. 60 t	Find site
Cape York, Greenland		Octahedrite		
Ahnighito (The Tent)	1894		30.9 t	New York
The Woman	1894		3 t	New York
The Dog	1894		400 kg	New York
Savik I	1913		3.4 t	Copenhagen
Agpalilik	1963		20 t	Copenhagen
Tunorput	1984		250 kg	Godthab

Table 9. Continued

Name	Date of fall or year of find	Class	Weight	Present location
Bacubirito, Mexico	1863	Octahedrite	ca.22 t	Culiacan
Armanty, China	1898	Octahedrite	20 t	Find site
Campo del Cielo, Argentina		Octahedrite		
El Mesón de Fierro	1576		15 t	Find site ?
Otumpa	1803		0.9 t	London
El Toba	1923		4.2 t	Buenos Aires
El Mataco	1937		1 t	Rosario
El Taco	1962		2 t	Washington
(No name)	1969		18 t	Find site
Mbosi, Tanzania	1930	Octahedrite	16 t	Find site
Willamette, USA	1902	Octahedrite	14 t	New York
Chupaderos, Mexico		Octahedrite		
Chupaderos I	1854		14 t	Mexico City
Chupaderos II	1854		6.8 t	Mexico City
Adargas	Before 1600		3.4 t	Mexico City
Morito, Mexico	Before 1600	Octahedrite	10.1 t	Mexico City

Fig. 58. The iron meteorite Hoba on its find site in Namibia. (After Schneiderhöhn, Centralbl Miner., 1931)

the three large iron blocks, and he brought them to New York by ship. Today, they are exhibited there in the American Museum of Natural History. In 1963 the Danish meteorite researcher Vagn Buchwald found a 20-t block in the same region that had not been known to the Eskimos. He succeeded without the help of heavy machinery in loading the specimen onto a ship and transporting it to Copenhagen (Fig. 59).

The history of the large iron meteorites from Campo del Cielo in Argentina is interesting. They had long been known by the local inhabitants who apparently knew about their origin, because they named the place where they were found *Campo del Cielo* = Field of the Heavens. The Spanish governor sent an expedition there in 1576 that found the 15-t iron block "Mesón de Fierro" (large table of iron). Later expeditions in the 18th century thought that it was a silver ore and tried to recover the silver. The enlightened Europeans of the time did not believe that these

Fig. 59. Transport of the 20-t iron meteorite Agapalilik (Cape York, Greenland), found 1963, on a sledge towards the coast. (From Buchwald, Handbook of Iron Meteorites, Univ. Calif. Press, 1975)

specimens could have fallen from the heavens. As they were not successful in recovering silver, the location of the "Mesón de Fierro" was forgotten and to date has not yet been relocated. However, many other specimens were found in the same area over the years (Table 9). The most important work, however, including the discovery of the associated crater field was done by an Argentine-American expedition after 1960 (Table 4).

The curious shape of the Willamette iron meteorite is remarkable (Fig. 60). On the base of this cone-shaped meteorite are a number of large, irregular, and roundish depressions in which small children can find comfortable seats. It is not completely clear whether these holes were formed by the melting of

Fig. 60. Iron meteorite Willamette, Oregon, USA. (After Hovey)

Fig. 61. Iron meteorite E1 Morito on display in Mexico City. (After Farrington)

low-melting iron sulfide in the atmosphere or, more probably, by weathering.

Chiseled in the Morito iron that is known since 1600 was the following text:

Solo Dios con su poder
Este fierro destruira
Porque en el mundo no habra
Quien lo puedo deshacer (1821)

106

This can be translated roughly into English: "Only God with his might can split this iron, as man has no means by which it can be divided into parts." Apparently, people had also tried in vain to break off metal pieces from this meteorite mass. Figure 61 shows this 10-t iron meteorite as it can be seen today in Mexico City in the Palazzio de Mineria, Tacuba No. 5.

2.3 The Shape of Meteorites

Meteorite shapes vary greatly. They are apparently fragments of accidental form as they enter the atmosphere. These may be shattered further during flight; on the other hand, ablative melting flattens and rounds them. Frequently conical, approximately pyramidal shapes are found. Figure 62 illustrates this with the North Chile (Buey Muerto) meteorite, and Fig. 63 with the Long Island, Kansas (USA), stony meteorite, the third largest stone known. Much rarer are elongated or columnar shapes such as the iron meteorite

Fig. 62. Iron meteorite North Chile (Buey Muerto)

Fig. 63. Stony meteorite Long Island, Kansas, USA. Length of the lower basis, 70 cm. (After Farrington)

Fig. 64. Iron meteorite Babb's Mill (Blake's Iron), Tennessee, USA, about 1 m long. (After Farrington)

Fig. 65. Iron meteorite Boogaldi, New South Wales, Australia. About 1/3 natural size. (After Liversidge, Proc. R. Soc. N.S. Wales, 1902)

a

b

Fig. 66a, b. The stony meteorite Krähenberg, Pfalz, Germany, which fell on May 5, 1869, weight 16 kg. **a** Front side (20 × 30 cm) with the regmaglypts radiating from the center; **b** side view, front side on top. (Photographs Historisches Museum der Pfalz, Speyer)

from Babb's Mill, Tennessee (USA) (Fig. 64). Irregular and jagged specimens have long been known from the Henbury, Australia, meteorite crater (cf. Fig. 21, p. 39), and from the Sikhote-Alin shower.

The smaller stony meteorites have predominantly irregular, roundish, lumpy shapes.

For many meteorites two distinct and differently formed sides may be discerned that are referred to as forward and rear. These are *oriented meteorites*. Since they evidently maintained a constant position during their passage through the atmosphere, the forward surface differs from the rear surface. Figure 65 illustrates the beautifully oriented iron meteorite from Boogaldi, New South Wales; the forward side is to the right. The meteorite North Chile (Fig. 62) also shows orientation, the point forms the forward surface. The oriented stone meteorite from Krähenberg in the Pfalz region of Germany, which fell in 1869, is shown in Fig. 66.

2.4 Surface Characteristics

Many meteorites show highly characteristic smooth, bowl-shaped depressions in their surfaces, which often look like thumb prints. They are easily recognized

Fig. 67. Front side of the iron meteorite Cabin Creek, Arkansas, USA, height 44 cm, weight 48 kg. (After Berwerth, Ann. Naturhist. Hofmuseum Wien, 1913)

Fig. 68. The chondrite Innisfree, Canada, showing the dark fusion crust and the gray interior; about half the natural size. (Photograph Smithsonian Institution)

on the stone from Krähenberg (Fig. 66), and the irons from Treysa (Fig. 57) and Cabin Creek (Fig. 67). They are called *regmaglypts* from the Greek words *regma* = cleft, and *glypto* = cut, i.e., features cut into the surface. They develop during the flight of the meteorite through the atmosphere. The turbulent air flow forms the depressions as a result of uneven melting of the surface. As a result of the compressed air flowing from the center outwardly, they are often oriented radially (see Fig. 66).

All meteorites that are found shortly after their fall and that have not been shattered on landing are enclosed in a black, mat, or shiny fusion crust that, at least for most meteorites, stands in sharp contrast to their light-colored interiors. This contrast is clear in Fig. 68. The crust is normally paper thin, less than 1 mm in thickness, but can be as thick as several

Fig. 69. Fusion crust of the achondrite Sioux County, USA, with flow structures, half the natural size. (Photograph Smithsonian Institution)

millimeters. On the forward side it is generally thinner than on the rear surface. It consists of dark, iron-containing glass in the case of the stone meteorites, and of iron oxides for the iron meteorites. Particularly for the iron meteorites, the crust is rapidly affected by weathering and converted into a skin of rusty iron oxides. That the crust is actually a product of melting is frequently demonstrated by flow structures that are well illustrated in Figs. 65 and 69. The molten material moves with the air flow as stringers to the rear of the specimen. At the edges, between the forward and rear surfaces, an accumulation of melt crust develops that may be several millimeters thick. The dark fusion crusts are an important distinguishing characteristic for meteorites. For light-colored stone meteorites that can appear very similar to terrestrial rocks, it provides an easily recognizable characteristic.

2.5 Mineralogy and Classification of Meteorites

We have already mentioned the main classes of meteorites. Now we will discuss individual classes in more detail and describe their constituent minerals (Tables 10 and 11). The meteoritic silicate minerals olivine, pyroxene, and feldspar are also important minerals found in the crust of the Earth, and both the meteoritic and terrestrial minerals have identical properties. But in addition, meteorites contain metallic iron and minerals that formed in an environment of very little oxygen and free of water. Most of the carbides, nitrides, and sulfides listed in Table 11 belong to this category. On the other hand, water- or hydroxyl-containing silicates (for example, mica or hornblende) are not known in meteorites with the exception of the carbonaceous chondrites.

2.5.1 Stone Meteorites

Stony meteorites consist predominantly of the silicate minerals olivine, pyroxene, and feldspar. They may also contain metallic iron and iron sulfide, but always in lesser amounts. The stony meteorites are further divided into *chondrites* and *achondrites*.

2.5.1.1 Chondrites

The characteristic components (40 to 90%) of chondrites are the *chondrules* or small spheres, ranging in diameter from 0.2 to 1 mm. On a broken surface they catch the eye because they protrude as hemispheres or because their color differentiates them from the fine-grained groundmass (see Fig. 53). Chondrules are easily removed from some chondrites (Fig. 70), while in others they are firmly incorporated into their surroundings. They consist chiefly of olivine and

Table 10. The main minerals in meteorites

Mineral	Formula	Occurrence[a]
Olivine, solid solution of		Ch, Pal, Mes, Ach, (Fe)
Forsterite	Mg_2SiO_4	
Fayalite	Fe_2SiO_4	
Orthopyroxene, solid solution of		Ch, Ach, Mes, (Pal), (Fe)
Enstatite	$MgSiO_3$	
Ferrosilite	$FeSiO_3$	
Clinopyroxene, solid solution of		Ch, Ach, Mes, (Fe)
Diopside	$CaMgSi_2O_6$	
Hedenbergite	$CaFeSi_2O_6$	
Augite	$(Ca, Na, Mg, Fe, Mn, Al, Ti)_2(Si, Al)_2O_6$	
Pigeonite	$(Mg, Fe, Ca)_2Si_2O_6$	
Feldspar, solid solution of		Ach, Ch, Mes, (Fe)
Anorthite	$CaAl_2Si_2O_8$	
Albite	$NaAlSi_3O_8$	
Orthoclase	$KAlSi_3O_8$	
Nickel-iron		Fe, Mes, Pal, Ch, (Ach)
Kamacite, α-iron	FeNi, 4–7% Ni	
Taenite, γ-iron	FeNi, 20–50% Ni	
Tetrataenite	FeNi, 50% Ni	

Mineral	Formula	Occurrence
Troilite	FeS	Ch, Fe, Mes, Pal, (Ach)
Clay Minerals Fe-serpentine Septechlorite Cronstedtite	Water-containing Silicates, main elements: Si, Fe, Mg, Al, Ca, ca. 10% water	cCh: CI and CM
Chromite	$FeCr_2O_4$	Ch, Ach, Mes, Pal, (Fe)
Magnetite	Fe_3O_4	cCh, Ch: type 3
Ilmenite	$FeTiO_3$	Ch, Ach, Mes
Spinel	$MgAl_2O_4$	cCh, Ch: type 3
Apatite	$Ca_5(PO_4)_3Cl$	Ch, Mes, (Fe)
Whitlockite	$Ca_3(PO_4)_2$	Ch, Ach, Mes, Pal, (Fe)
Pentlandite	$(Fe,Ni)_9S_8$	cCh, Pal
Schreibersite	$(Fe,Ni)_3P$	Fe, Mes, Pal
Cohenite	Fe_3C	Fe, ECh, (Ch: type 3)

---

[a] Occurrence: Ch = chondrites, Ach = achondrites, cCh = carbonaceous chondrites, ECh = enstatite chondrites, Mes = mesosiderites, Pal = pallasites, Fe = iron meteorites, CAI = Ca,Al-rich inclusions. ( ) = Rare in this class.

Table 11. Accessory minerals in meteorites

Mineral[a]	Formula	Occurrence[b]
Elements, carbides, nitrides, silicides:		
Diamond	C	Ach: ureilites; Fe, cCh
Graphite	C	Fe, Ch, Mes, Ach: ureilites
Copper	Cu	Ch
*Haxonite	$(Fe,Ni)_{23}C_6$	Fe, Ch: type 3
*Carlsbergite	CrN	Fe
*Osbornite	TiN	Ach: aubrites
*Sinoite	Si_2N_2O	ECh
*Perryite	$(Ni,Fe)_2(Si,P)$	Fe
Sulfides:		
Alabandite	$(Mn,Fe)S$	ECh, Ach: aubrites
*Daubréelite	$FeCr_2S_4$	Fe, ECh
Djerfisherite	$K_3Cu(Fe,Ni)_{12}S_{14}$	ECh
*Heideite	$(Fe,Cr)_{1+x}(TiFe)_2S_4$	Ach: aubrites
*Niningerite	MgS	ECh
*Oldhamite	CaS	ECh, Ach: aubrites
Oxides:		
Hibonite	$CaAl_{12}O_{19}$	cCh: CAI
Perovskite	$CaTiO_3$	cCh: CAI
Rutile	TiO_2	Mes, Ch

Carbonates, sulfates:

Calcite	$CaCO_3$	cCh: CI, CM
Dolomite	$CaMg(CO_3)_2$	cCh: CI
Magnesite	$(Mg,Fe)CO_3$	cCh: CI
Epsomite	$MgSO_4 \cdot 7H_2O$	cCh: CI
Gipsum	$CaSO_4 \cdot 2H_2O$	cCh: CI

Phosphates:

*Farringtonite	$Mg_3(PO_4)_2$	Pal
*Panethite	$(Na,Ca)_2(Mg,Fe)_2(PO_4)_2$	Fe
*Brianite	$Na_2CaMg(PO_4)_2$	Fe, cCh
*Stanfieldite	$Ca_4(Mg,Fe)_5(PO_4)_6$	Pal, Mes

Silicates:

Quartz	SiO_2	Ach, ECh
Cristobalite	SiO_2	ECh
Tridymite	SiO_2	Ach, ECh, Mes
*Ringwoodite	$(Mg,Fe)_2SiO_4$	Ch: in shock veins
Melilite	$Ca_2(Al,Mg)(Si,Al)_2O_7$	cCh: CAI
Nepheline	$KNa_3(AlSiO_4)_4$	cCh: CAI
Sodalite	$Na_8(AlSiO_4)_6Cl_2$	cCh: CAI

a*, Not known from terrestrial rocks.
b Abbreviations, see Table 10.

Fig. 70. Chondrules separated from the L-chondrite Saratov (grid 1 mm)

pyroxene, with a small amount of feldspar that is generally found as a binding agent filling the spaces between the olivine and pyroxene crystals. Their structure becomes clear when examined in thin section under the microscope. Figure 71 illustrates several chracteristic chondrule types.

Chondrites also contain metallic nickel–iron in irregularly shaped, millimeter-sized grains. On a cut slice they stand out prominently as bright, reflecting grains (Fig. 72). On the ground they develop a halo of brown rust fairly rapidly. Additional constituents of chondrites are the iron sulfide mineral, troilite (about 5%); chromite and apatite in smaller amounts; and a few rarer minerals (see Tables 10 and 11).

Chrondrules, broken pieces of chondrules, nickel-iron, and other mineral grains are embedded in a fine-grained groundmass (grain size smaller than 0.1 mm), the matrix. It consists essentially of the same minerals as the larger components, but contains additionally very fine materials (iron-rich olivine, feldspar, and nepheline) that are rich in volatile elements and carbon. The whole assemblage is an *undifferentiated* conglomerate of high- and low-temperature minerals (chondrules and matrix), and light and heavy components (silicates and metal/sulfide). This assemblage

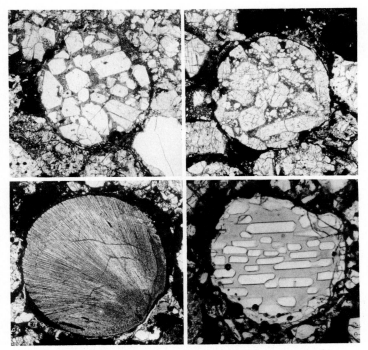

Fig. 71. Chondrule types as seen through the microscope in thin section. *Upper row* Most common are porphyritic chondrules with olivine and/or pyroxene crystals in a glassy or fine-grained groundmass. *Lower left* Excentro-radial pyroxene chondrule. *Lower right* Barred olivine chondrule. Side length of all pictures, 1 mm

is called undifferentiated because such a mixture is transformed upon heating and becomes differentiated into new components. The matrix becomes coarsely crystalline; metal and sulfide melt and sink because of their greater density; chondrules partially melt and feldspar separates from olivine and pyroxene. Chondrites, therefore, represent *primary material* that has never been melted as a whole.

Fig. 72. Cut surface of the LL chondrite Parnallee. The bright grains are metallic nickel-iron. The rounded, light, or dark structures are cross-sections of chondrules (two times natural size)

Fig. 73. Cut surface of the chondrite Paragould, length 27 cm. Brecciated texture with light clasts in a darker groundmass. (Photograph Smithsonian Institution)

The broken surface of a chondrite is normally light gray and can range to dark gray and black, black being particularly associated with the carbonaceous chondrites. Often chondrites consist of different colored fragments (breccias). They then appear mottled, or show on broken surfaces lighter pieces in a darker groundmass (Fig. 73). Darker material may also run through the stone as veins. Chondrites will now be further divided according to their chemical composition and structure.

2.5.1.1.1 Ordinary Chondrites

The name ordinary chondrite (OC) is given to the group that is most abundant. They are classified on the basis of total iron content and the iron content in their olivines and pyroxenes into groups designated as H-, L-, and LL-chondrites (Table 12). H stands for *high total iron*, L for *low total iron*, and LL for *low total iron* and *low metal*. Figure 74 shows the relationship between the content of iron metal and iron oxide: H-chondrites with the higher total iron content have a higher fraction of metal, but less oxidized iron; L- and LL-chondrites have a lower content in total iron, but more iron oxide and less metal. This relationship has been long known as *Prior's rule*: the more oxidized iron a chondrite contains, the less iron metal it has, and the richer that metal is in nickel. The reference to the nickel content results from the fact that chondrites have an essentially constant nickel content, and all of the nickel is found in the metal. The oxidized iron is essentially contained in the olivine and pyroxene, thus the Fe/(Fe + Mg) ratio in these minerals increases from H to L to LL (Table 12).

Table 12. Chemical classes of chondrites

Class	Old name	Total iron (wt%)	Metal (wt%)	Fa in Olivine (mol%)
Enstatite chondrites		22–33	17–23	Below 1
H-chondrites	Bronzite chondrites	25–30	15–19	16–20
L-chondrites	Hypersthene chondrites	20–24	4–9	22–25.7
LL-chondrites	Amphoterites	19–22	0.3–3	26.6–32
Carbonaceous chondrites		19–26	0–5	0–40

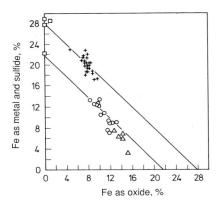

Fig. 74. The relation between iron as oxide and as iron metal and sulfide in enstatite chondrites (□), H-chondrites (+), L-chondrites (○) and LL-chondrites (△), after chemical analyses by A.J. Easton, E. Jarosewich and B. Wiik. The *diagonal lines* correspond to a constant total iron content of 27.7% (H-chondrites) and 21.5% (L-chondrites), respectively

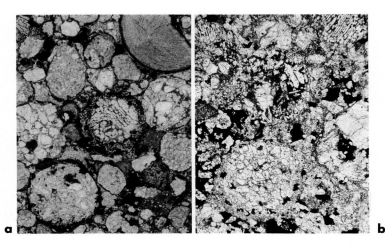

Fig. 75. Chondrites of different petrologic type (thin sections, transmitted light). **a** Tieschitz, type 3; **b** Simmern, type 5. Type 3 shows clearly delineated chondrules, in type 5 they appear blurred by recrystallization (small side of picture, 3 mm)

Table 13. Petrologic types of chondrites

	1	2	3	4	5	6
Texture	No chondrules	Chondrules very sharply defined		Well defined	Readily delineated	Poorly defined
Matrix	Finegrained, opaque	Chiefly fine, opaque	Clastic, opaque	Fine ----------> recrystallized, transparent		
Olivine, pyroxene composition		> 5% Mean variation of iron		Up to 5% variation	Homogeneous	
Low-Ca pyroxene		Chiefly monoclinic		Partly monoclinic	Orthorhombic	
Secondary feldspar		Absent		Fine-grained	Medium <50 µm	Coarse >100 µm
Chondrule glass		Clear, isotropic		Devitrified	Absent	
Sulfides		> 0.5% Ni		<0.5% Ni		
Metal		No taenite	Kamacite and taenite			
Water, wt%	ca.20%	4–18%	0.3–3%	Less than 1.5%		

The H-, L- and LL-chondrites are also referred to as *chemical groups*. These are then further divided into *petrographic types* that are chemically similar, but differ petrographically and are designated by the Arabic numbers 1 to 6. Types 1 and 2 apply only to the carbonaceous chondrites (see below). Type 3 chondrites are *unequilibrated*, i.e., their constitutent minerals are of variable composition and not constant, as is required for chemical equilibrium. Most important here is the variable composition of the olivine and pyroxene. In the *equilibrated* chondrites, types 4 to 6, iron and magnesium are evenly distributed within individual minerals; each grain of olivine has its appropriate equilibrium iron content as does each pyroxene grain. Structural differences are also observed; in type 3 chondrites the chondrules stand out clearly from the groundmass (mostly darker). For types 4–6 crystallization increases, the chondrules grow together with the coarsening groundmass and with each other, so that the chondritic structure becomes increasingly indistinct (Fig. 75). Ordinary chondrites are thus designated by abbreviations for the chemical class (H, L, LL) and petrographic type (3–6); for example, the chondrite Mainz is classified as an L6 chondrite.

Table 13 illustrates the distinguishing characteristics of the petrographic types as well as more subtle mineralogical differences between types 4, 5, and 6.

2.5.1.1.2 Enstatite Chondrites

Enstatite chondrites have about the same total iron content as H-chondrites (Fig. 74), but all of their iron is in the form of metal or sulfide and practically no iron oxide is found in the silicates. Compared with the ordinary chondrites they represent a recognizably separate group also because they contain many unusual minerals and differ in their trace element content. These minerals differ from those in ordinary chondrites in that they are formed in environments significantly poorer in oxygen. As a consequence, all of the iron and even some silicon is reduced to metal. Sulfur also combines with metals that would normally be found in combination with oxygen: magnesium, manganese, titanium, chromium (see Table 11).

The enstatite chondrites can be divided into two distinct groups, a high iron group, EH, with higher contents in iron and sulfur (about 30% total iron, 5.5% S), and a low iron group, EL, with lower concentrations (25% total iron, 3.5% S). The differences extend also to trace elements contained in metal and sulfides, and to the minor element content of the minerals. The known members of the EH group belong to petrographic types 3, 4, and 5, whereas all EL members are type 6.

2.5.1.1.3 Carbonaceous Chondrites

Carbonaceous chondrites are designated by the symbol C and were formerly divided into three types: C1, C2, and C3. Their name is derived from their black color, and those of type 1 are particularly structureless, similar to a piece of coal. C1 chondrites consist essentially of a fine-grained matrix and contain no chondrules (Fig. 76); nevertheless, they belong to the chondrites because they have the same major element composition. The other carbonaceous chondrites have a high matrix/chondrule ratio, usually larger than or equal to one. This groundmass is rich in water in the form of

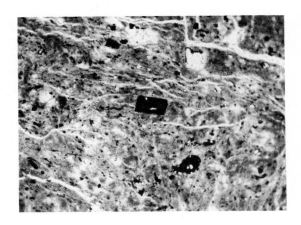

Fig. 76. Structure of the CI chondrite Orgueil (thin section, transmitted light, area 0.6 × 0.4 mm). The dark rectangular grain in the center is iron sulfide

hydrated minerals, and rich in sulfur, carbon, and organic compounds. In contrast to the group of OCs, the chemical and mineralogical composition changes when going from type C1 to C2 and C3. Therefore, it is better to use a different nomenclature to avoid misleading analogies to the OCs. Thus, CI (formerly C1), CM (formerly C2), CO and CV (both formerly C3) are distinguished, and as a new group, CK chondrites. Members of these groups can also be assigned petrographic types, e.g., CV3 or CK4.

The most primitive in the chemical sense (see Sect. 3.3) are the CI chondrites, named after the type specimen Ivuna. They consist of up to 99% of a fine-grained mixture of hydrated phyllosilicates (see Table 10). The rest is made up of grains of pyrrhotite, magnetite, and occasionally larger crystals of olivine and pyroxene. Carbonates [Ca-carbonate, breunerite (Fe, Mg) $CaCO_3$, dolomite $CaMgCO_3$] and sulfates of Na, Mg, Ca, and Ni also occur in the form of isolated grains or in veins cutting through the matrix. These were probably formed by aqueous activity on a parent body that did not change the bulk meteorite composition (see Sect. 3.3).

Unfortunately, only five representatives of these CI chondrites are known. The largest is Orgueil, which fell in France 14 May 1864. It was studied soon after recovery, disclosing its organic material. During the 1960s, there was a scientific controversy over peculiar structures found in Orgueil, the so-called *organized elements* that were held as fossils, the remains of living organisms. They were about 0.01 mm in size, six-sided or roundish in form, frequently decorated with regular, ordered, small projections that were reminiscent of pollen or single-celled life-forms. It was proven, however, that they were in part inorganic crystals and in part terrestrial contamination that included pollen. The significance of organic substances in carbonaceous chondrites will be treated in Section 2.9.

The CM chondrites (after the type specimen Mighei) consist of about 50% of a hydrous, fine-grained groundmass similar to CI chondrites. The other half is composed of high-temperature components: chondrules, aggregates, and crystal fragments. The chondrules are similar to the chondrules in OCs in that they are apparently frozen droplets of silicate melts, crystallized olivine and pyroxene in a glassy, feldspathic mesostasis. The aggregates, however, consist of loosely packed crystals that lack a magmatic texture. They are composed of olivine and pyroxene (mostly forsterite and enstatite, i.e., Fe-free), Ca, Al-rich glass, metal, and sometimes also of refractory minerals like hibonite, perovskite, and spinel.

The C3 chondrites are now separated into two groups: CO (after the meteorite Ornans) and CV (after the meteorite Vigarano), which differ in chemical and mineralogical composition. They belong mainly to petrologic type 3, except some CO2s and CV4s, e.g., Coolidge. Only the COs contain hydrated silicates in their matrix like the CMs, and the main matrix components are small Fe-rich olivines and Fe-sulfides, which alone make up the matrix of CV chondrites. Their content in carbon and organic compounds is lower than in the CMs, usually below 1% C. They differ petrographically in chondrule size and abundance: CO chondrites have small chondrules (0.1–0.4 mm) and less matrix (about 34%) than CVs, where the chondrules range from 0.2 to several millimeters, and the matrix makes up 42%. There are also chemical differences, they are discussed in Section 2.7. Both types contain *refractory inclusions*, but they are more prominent in CV chondrites like Allende. They appear as light colored, almost white inclusions that can be as large as several centimeters on broken surfaces of CV chondrites (Fig. 77). They are Ca- and Al-rich and consist of minerals that are otherwise not found in chondrites: melilite, Ti-rich pyroxene (fassaite), Na-free anorthite, Fe-free spinel, hibonite, perovskite, and platinum- and iridium-rich metal. These are without exception high-temperature minerals that were most likely formed as the first condensation products created as the planetary system formed from the solar nebula (see below). Certainly, the carbonaceous chondrites represent the most primitive meteoritic material known, material that has changed very little since its formation.

Recent finds in Antarctica have helped to establish an additional group: the CK chondrites. Six

Fig. 77. Slice of the carbonaceous chondrite Allende. In the dark matrix numerous light chondrules and several large, white inclusions are visible

Antarctic finds, and one find (Maralinga) and one fall from Australia (type specimen Karoonda) make up this group. All members are equilibrated and belong to petrographic types 4 to 6, except a recent find from the Australian Nullarbor, which is the first CK3. They are highly oxidized and contain no metal; magnetite and pentlandite are the main opaque phases; pyrrhotite, pyrite, and sulfides of noble metals are found as minor phases. Iron-rich silicates [olivine Fe/(Fe + Mg) 29–33 mol %, pyroxene Fe/(Fe + Mg) 23–29 mol %] occur in chondrules and as crystal fragments in the anhydrous groundmass, which consists mainly of olivine. The bulk chemical composition is similar to, but different in detail, from the CO and CV groups.

There remain several ungrouped carbonaceous chondrites, which do not fit into any of the established groups. The main criterion for including a chondrite into the carbonaceous clan is not its carbon content, but its bulk chemistry: an enrichment of *refractory lithophile* elements (Ca, Al, Sc, etc.) over the level in CI. This will be explained in more detail in Section 2.7.

2.5.1.2 Achondrites

Achondrites contain no chondrules, but their differences to chondrites are more far-reaching: achondrites are *differentiated* meteorites, i.e., they derived from primary, more primitive material by a melting process. They were chemically changed during this process, in contrast to the chondrites, which were not differentiated and therefore retain their original composition. The achondrites are composed essentially of the minerals pyroxene, feldspar, and olivine; the minerals that also make up most of the rocks of the Earth's crust. Their proportions and compositions are highly variable (Table 14). Their texture is magmatic as their minerals were separated from a liquid melt (Fig. 78). These rocks are frequently fragmental, having been reassembled into breccias consisting of a fine-grained matrix containing various types of larger fragments.

2.5.1.2.1 Eucrites, Howardites, and Diogenites

These are the most abundant achondrites and resemble terrestrial basalts in their mineralogy. Consequently, they are difficult to distinguish from terrestrial rocks and are seldom recognized as finds. They are

Table 14. Composition of achondrites

Class	Pyroxene Content (%)	Pyroxene Fs (mol%)	Feldspar Content (%)	Feldspar An (mol%)	Olivine content (%)
Eucrites	40–65	40–70	30–55	85–95	Absent
Howardites		(Breccias from eucrites and diogenites)			
Diogenites	90	25	ca. 2	87	Rare
Hypersthene-achondrites	90	25	ca. 2	87	Rare
Shergottites	70	20–60	20	43–57	Rare
Pigeonite-maskelynite-achondrites	70	20–60	20	43–57	Rare
Nakhlites	80	40	Minor	23–36	5–10
Augite-olivine-achondrites	80	40	Minor	23–36	5–10
Chassignites	5–8	12–28	2–8	16–37	80–90
Olivine-achondrites	5–8	12–28	2–8	16–37	80–90
Aubrites	97	0	Rare	–	Absent
Enstatite-achondrites	97	0	Rare	–	Absent
Ureilites	20–50	5–25	Absent	–	40–80
Olivine-pigeonite-achondrites	20–50	5–25	Absent	–	40–80

Fs = mol% Fe/(Fe + Mg); An = mol% anorthite.

Fig. 78. Cut surface of the achondrite Shergotty, natural size. (Photograph Smithsonian Institution)

recovered only when their fall was observed, or under the favorable collecting conditions of Antarctica or barren deserts like the Nullarbor in Australia. They consist of pyroxene and feldspar with small amounts of silica (quartz or tridymite), phosphate, chromite, and iron sulfide. In contrast to terrestrial basalts (disregarding the mentioned exceptions), they also contain some metallic iron (0.1–1%), much less than is present in chondrites. Diogenites contain mainly only pyroxene that is poorer in iron than the pyroxene in eucrites. Howardites are mineralogical and compositional intermediates between these two types. They are breccias of rock and mineral fragments of eucrites and diogenites that probably formed by impact on the surface of their parent body.

2.5.1.2.2 SNC Meteorites

This acronym brings together three types of achondrites: shergottites, nakhlites, and chassignites. The shergottites are similar to the eucrites, consisting primarily of pyroxene and feldspar. The feldspar, however, is not crystalline but has been transformed to an isotropic glass (maskelynite), probably by shock waves. It is also more sodium-rich than the feldspar of the eucrite group. Nakhlites and chassignites also contain a soduium-rich feldspar, but it has not been transformed to a glass. None of the three classes contain metallic iron, but some may contain minerals with crystallization water such as hornblende or iddingsite. We infer from this that they formed under more oxidizing conditions than the eucrites. A further characteristic of the SNC meteorites is their young crystallization ages. While for the chondrites and the eucrite group, ages of 4.5 billion years have been determined, the SNC meteorites are only a few hundred million years old. This suggests that their origin was in a relatively large body, which underwent magmatic and volcanic activity that continued long after its formation (similar to occurrences on the Earth). Further chemical evidence has led to the suggestion that this body could be the planet Mars (see Sect. 3.2.2).

2.5.1.2.3 Aubrites

Aubrites are also referred to as enstatite achondrites, because they consist almost completely of iron-free pyroxene, enstatite. They are comparable, therefore, to the enstatite chondrites, and contain as they do the unusual minerals that form under strongly reducing conditions, for example, CaS, MnS, and TiN. In contrast, however, they have a magmatic texture and are somewhat different chemically. It is assumed that they derived by melting and differentiation processes from the enstatite chondrites. Most of the aubrites were brecciated at a later stage.

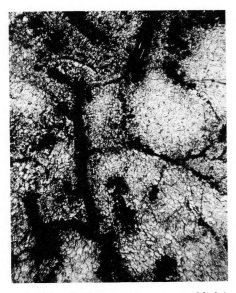

Fig. 79. Structure of the ureilite Haverö (thin section, transmitted light). Light areas are olivine crystals, transected by dark carbon veins. Small side of picture, 2 mm (After Wlotzka, Meteoritics 7, 1972)

2.5.1.2.4 Ureilites

Ureilites are distinguished from all other achondrites on the basis of their high carbon content (ca. 2%) and the occurrence of diamonds. The carbon occurs as veins less than 1 mm wide consisting of finely intergrown grains of graphite, nickel–iron, and troilite; they also contain tiny diamonds (smaller than 1 μm in size). These veins penetrate aggregates of olivine crystals with varying amounts of pyroxene (pigeonite, see Fig. 79). The texture is that of a cumulate of silicate grains which have collected on the floor of a magma chamber. The graphite veins appear as if they were introduced into the silicates later by a shock event,

135

one that could also account for the diamond content. Detailed studies have shown, however, that the graphite was already present in the magma chamber where the ureilites crystallized. The high carbon content along with the content of several trace elements indicates that ureilites could have originated from or together with the carbonaceous chondrites. The name *ureilite* was coined after the meteorite Novo-Urei from Russia.

2.5.2 Iron Meteorites

Iron meteorites are differentiated meteorites consisting of more than 90% metallic nickel-iron and only a few other minerals. Frequently, these minor or trace minerals occur in rounded nodules that consist primarily of troilite and/or graphite, often surrounded by schreibersite and cohenite (see Fig. 81b). Some iron meteorites also contain sizable silicate inclusions (see Fig. 89).

The classification of iron meteorites is based on their structure. To examine the structure of a specimen properly, a flat surface is prepared, ground smooth, and polished to a mirror finish. On this highly reflecting surface one can now distinguish only the included mineral bodies by their reflection colors, for example, the shiny yellow appearances of troilite and schreibersite. Upon etching the surface with acid (recipe in Appendix) various structures are revealed in the metal.

One group of iron meteorites reveals assemblages of very fine parallel lines that cross each other at various angles and are referred to as *Neumann lines* after their discoverer. Occasionally, such a system of

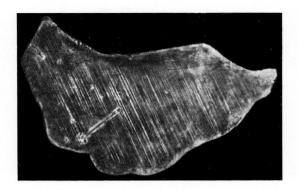

Fig. 80. Neumann lines in the iron meteorite North Chile (Buey Muerto), etched (1/4 natural size)

lines is revealed with particular clarity as in the North Chile (Buey Muerto) meteorite (Fig. 80). These lines are actually cross sections of very thin plates, so-called twin lamellae. They form because of mechanical stress, and for some iron meteorites it could be determined that they formed when the meteorite hit the ground. The iron meteorites whose structures are dominated by Neumann lines break easily into bodies with three planes perpendicular to each other. Because of this cleavage into hexahedrons, they are known as *hexahedrites*. They consist chemically of a uniform nickel alloy, the mineral kamacite (see below).

More frequently, etched iron meteorites display a more conspicuous structure. Again, one sees parallel assemblages of lamellae that intersect one another at various angles. They are usually much wider than the Neumann lines. This characteristic structure of interpenetrating lamellae is called a *Widmanstätten pattern* after its discoverer (Fig. 81). Examination of this

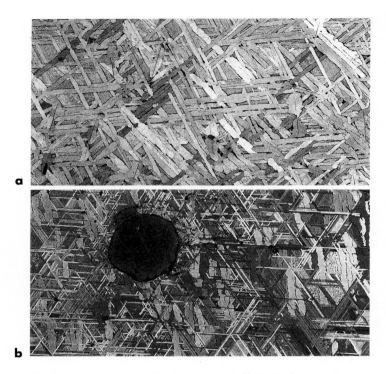

Fig. 81. Polished and etched surface of octahedrites, natural size. **a** Coarse octahedrite Staunton, lamellae 1.6 mm; **b** fine octahedrite Altonah, lamellae 0.3 mm. On the *left* a troilite nodule with some graphitic carbon around the edges. (Photographs Smithsonian Institution)

structure with a magnifying glass reveals two major structural components. The dominating lamellae consist of the nickel-poor, iron-nickel alloy, *kamacite* (after the Greek word for beam). The kamacite lamellae are separated by much thinner bands of highly reflecting nickel-rich *taenite* (after the Greek work for ribbon). Figure 82 shows a microscopic view that clearly distinguishes these two structural elements. A third element,

Fig. 82. Widmanstätten pattern as seen through the microscope in the finest octahedrite EET 84300. The *broad bands* are kamacite (width about 0.1 mm), the *thin ribbons* taenite. (Photograph R.S. Clarke)

the material filling the wedges between the lamellae, is called *plessite* (after the Greek work for filling). It consists of a fine intergrowth of kamacite and taenite.

The clear resolution of the pattern by etching is a result of the differing acid resistance of the two nickel–iron alloys. The resistant taenite plates project above the surface as fine ridges, because the kamacite bands have dissolved more readily. When one etches long enough, the relief becomes deep enough so that an etched meteorite plate can be used as a printing block. This technique, known as *Naturselbstdruck* (nature printing), was devised and used by A. von Widmanstätten about 1830 before the development of photography. It was actually superior to modern reproduction processes, because with some magnification

Fig. 83. Nature print from the medium octahedrite Lenarto, threefold enlarged, showing the fine structure of plessite fields. (After Partsch, Die Meteoriten, Wien 1843)

the finest details could be recognized (Fig. 83). Today, this detail is lost due to the printing screens used.

How does the regular pattern of lamellae come about? Closer investigation shows that these iron meteorites are made up of plates of nickel-iron whose cross sections are presented on the cut surface. These plates consist of kamacite that are sheathed by thin films of taenite. They are arranged parallel to the four surface pairs of an octahedron. An octahedron is shown in Fig. 84, a double pyramid consisting of eight equilateral triangles. Now if an octahedron consisting of such plates is cut, various patterns result depending on the direction of the cut (Fig. 84). This orientation of lamellae is often uniform even in very large metal masses, showing that the structure developed from a single, homogeneous crystal.

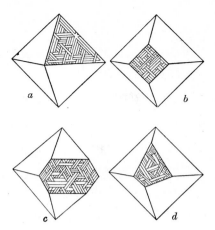

Fig. 84. Widmanstätten pattern as seen in different cuts through an iron meteorite. **a** Cut parallel to an octahedral plane; **b** cut parallel to a cubic plane; **c** cut parallel to a rhombic dodecahedral plane; **d** random cut. (After Tschermak, Lehrbuch der Mineralogie, 1894)

Iron meteorites that display these octahedral structures are called *octahedrites*. They are further subdivided into *coarse* to *fine octahedrites* based on the width of their kamacite lamellae (see Table 15).

The formation of the octahedral structure was a puzzle for many years, because it was not possible to produce similar structures in the laboratory. The explanation resulted from the detailed study of the iron (Fe) and nickel (Ni) alloy system at temperatures below the melting point. It is observed that these homogeneous alloys, consisting of taenite, disintegrate at 900°C upon slow, uninterrupted cooling into kamacite poorer in Ni and taenite richer in Ni than the starting materials. At a given temperature and at equilibrium both minerals have definite Ni contents that are given by the Fe-Ni equilibrium phase diagram (Fig. 85). The diagram shows how an alloy of 10% Ni behaves on cooling. Above 700°C only taenite is present; as this temperature is reached, small crystals of kamacite containing 3% Ni precipitate (point A). With further cooling the kamacite and

Table 15. Structural and chemical classes of iron meteorites

Structural class	Symbol	Bandwidth (mm)	Ni (%)	Chemical class
Hexahedrites	H	–	5–6	IIA
Coarsest octahedrites	Ogg	> 3.3	5–9	IIB
Coarse octahedrites	Og	1.3–3.3	6.5–8.5	IAB, IIIE
Medium octahedrites	Om	0.5–1.3	7–13	IID, IIIAB
Fine octahedrites	Of	0.2–0.5	7.5–13	IIIC, IVA
Finest octahedrites	Off	< 0.2	17–18	IIID
Plessitic octahedrites	Opl	< 0.2, Spindles	9–18	IIC
Ataxites	D	–	16–30	IVB

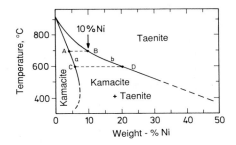

Fig. 85. Iron–nickel phase diagram, (After Goldstein and Ogilvie, Trans. Met. Soc. AIME 233, 1965)

taenite compositions must change according to curves a and b, whereby both become richer in Ni. At 600°C kamacite contains 6% Ni (point C) and taenite 20% (point D). Ni diffuses into both phases, which is only possible in a system of constant composition when part of the taenite dissolves and frees Ni, resulting in kamacite growth at the expense of taenite. With very slow cooling platy crystals of kamacite form and orient themselves along the octahedral planes of the original taenite crystal lattice. This growth process is the origin of the Widmanstätten structure.

It is possible through precise analyses of the Ni distribution between kamacite and taenite to determine the cooling rate. The diffusion of Ni becomes slower with decreasing temperature. Finally, taenite can no longer adjust its interior Ni content, it can only increase it to the high Ni value required at low temperatures at its boundary with kamacite. The result is the characteristic M-shaped Ni concentration profile (Fig. 86). If one knows the diffusion rates of Ni in kamacite and in taenite, one can determine the cooling rate of an iron meteorite from the analysis of the Ni-concentration profiles. Observed cooling rates range between approximately 1 and 100°C per million years for the temperature range from 700 to 450°C. At 350°C the rate is so slow that kamacite and taenite are

143

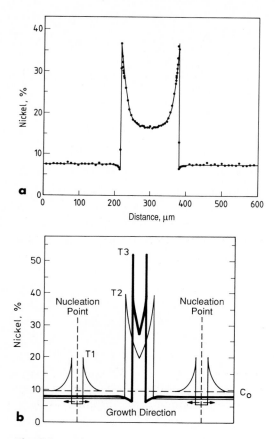

Fig. 86. a Ni distribution as measured in a taenite lamella of an iron meteorite; **b** schematic development of Ni distribution during cooling from temperature $T1$ to $T2$ to $T3$. (After Goldstein and Axon, Naturwissenschaften. 60, 1973)

no longer affected. This slow cooling over millions of years makes it understandable that this structure could not be reproduced in the laboratory.

Fig. 87. Iron meteorite Toluca, Mexico. *Above* Etched slice with Widmanstätten pattern. *Below* After heating to 950°C. About 1.5-fold enlarged. (After Berwerth, Sitz.Ber. Akad. Wiss. Wien, 1905)

Reheating of the structure to 900 to 1000°C for a short time, however, can destroy it. Here, we again refer to the phase diagram which indicates that at this temperature we are in the taenite field. During rapid cooling, this taenite cannot exsolve kamacite, but transforms into a matastable phase, called α_2. Upon etching an irregular mottled surface is revealed, replacing the Widmanstätten pattern (Fig. 87). Such secondarily modified structures are relatively common among iron meteorites, many of which have experienced a blacksmith's furnace or have been abused in some similar way. Buchwald found that 18% of all iron meteorites show such modifications. The 1.5-t iron meteorite from Bitburg in the Eifel (ca. 1805) was brought from its site of recovery to a blacksmith where an attempt was made to produce workable iron.

The attempt failed and the remaining pieces are useless to science.

During atmospheric flight, a thin surface zone of iron meteorites (ca. 1 cm deep) is altered by the heat. A granular α_2 structure may form to a depth of 2 mm. Beneath this, to a depth of 1 cm, changes due to heat can be demonstrated by hardness measurements. Further inside the meteorite no evidence of heat effects can be found.

The nickel content of iron meteorites ranges approximately between 5 and 35%. There are only a few exceptions with higher Ni, the highest one being Oktibbeha County with 60%. The Fe-Ni phase diagram (Fig. 85) shows that the structure is determined by the Ni content: between 6 and 20% octahedrites form, and within this range the higher the Ni content, the finer the resulting kamacite bands (and the larger the amount of residual taenite). These observed differences have led to the classification of octahedrites based on the width of their kamacite bands (see Table 15 and Fig. 81).

With Ni contents of less the 6%, essentially all of the taenite converts to kamacite, resulting in the previously mentioned hexahedrites. With slightly higher Ni we have the transition to the coarsest octahedrites, i.e., those meteorites that are made up of only a few broad kamacite lamellae and very little taenite. With more than 20% Ni, the formation of kamacite becomes increasingly difficult as the temperature at which kamacite precipitates decreases and diffusion becomes slower and slower. The result is the formation of the finest octahedrites and plessitic octahedrites, and finally the structureless *ataxites* (Greek = without structure). There are also meteorites that appear similar to ataxites, but which contain much less Ni. Examples are the Rafrüti iron (Canton of Bern, Switzerland, found in 1886) with 9.4% Ni, or the meteorite that fell in 1968 near Juromenha (Alentejo) in Portugal with 8.7% Ni. These iron meteorites consist of essentially fine-grained kamacite that undoubtedly formed when an octahedrite was reheated in the cosmos and then rapidly cooled.

In addition to the divisions based on structure, a chemical classification has also been developed. This new approach has now become the definitive classification system for iron meteorites because of its success in determining geochemical relationships between various meteorite groups. It uses nickel and the trace elements gallium, germanium, and iridium to define chemical groups. These trace elements show correlations with nickel, and in diagrams such as Fig. 88, related compositions are grouped in fields. The iron meteorites that cluster in these various groups are designated by Roman numerals and letters (IAB, IIA, etc.). Usually, the various structural classes belong to specific chemical groups. Table 15 lists the structural classes and their corresponding chemical groups. It can be assumed that iron meteorites in a given chemical group have the same origin and come from the same parent body.

Iron meteorites may also contain silicate inclusions. They occur most commonly in the coarse octahedrites of group IAB and may form centimeter-sized aggregates (Fig. 89). These consist of approximately 0.1-mm crystals of olivine, pyroxene, and feldspar and are similar in composition to the chondrites. They are distinct from the pallasites that contain only the silicate olivine, normally in large crystals. The origin of these chondritic inclusions and their relationship to the chondrites, however, are still not clear.

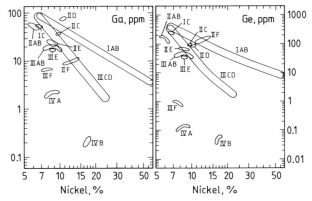

Fig. 88. Gallium–nickel and germanium–nickel diagram with the areas for the different chemical groups of iron meteorites. (After Wai and Wasson, Nature 282, 1979)

Fig. 89. Silicate inclusions (*dark*) in the octahedrite Landes. Length of picture, 3 cm. (After Luzius-Lange, Diss. Mainz, 1986)

The Steinbach meteorite (found in the Erzgebirge in 1724) probably also belongs to the group of iron meteorites. It is made up of about equal parts of nickel-iron and silicates that have grown together in a coarse-grained pallasite-like structure. The silicates are ortho- and clinopyroxene as well as tridymite. Earlier it was classified as a *siderophyre*. Its metal, however, displays a Widmanstätten pattern and its chemical composition places it in group IVA.

2.5.3 Stony Iron Meteorites

2.5.3.1 Pallasites

Pallasites were named after the explorer Peter Simon Pallas. In 1772, during travels in Russia, he studied and described a large iron mass that had been found on a mountain near Krasnojarsk. It contained cavities filled with large olivine crystals; this characteristic pallasite structure is shown in Fig. 90. The olivines are typically 0.5 to 2 cm across. When the metal surrounding these crystals is dissolved, many of them show crystal facets in which the edges between them are rounded (Fig. 91).

On the boundary between metal and olivine small amounts of other minerals are found: troilite, schreibersite, chromite, orthopyroxene, and phosphates. The metal consists of a Widmanstätten intergrowth of kamacite and taenite and may have large olivine-free domains (as in occasional specimens of the Brenham pallasite). The Ni and trace element content of the metal is similar to that of the IIIAB medium octahedrites. The pallasites presumably originate from

Fig. 90. Polished slice of the pallasite Admire. Olivine crystals appear dark in light metallic iron; approx. natural size

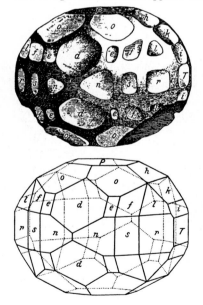

Fig. 91. Olivine grain with crystal faces from the pallasite Krasnojarsk. The *lower figure* shows the reconstruction of the crystal faces. About tenfold enlarged. (From Kokscharow, Mater. Mineral. Russlands, 6)

the boundary zone between the iron core and the silicate mantle of a smaller planet (asteroid).

2.5.3.2 Mesosiderites

Mesosiderites also consist of silicates and metal (40 to 60%). Their textures, however, are much finer grained and frequently very irregular. In metal-rich areas large silicate nodules may be enclosed (Fig. 92). The silicate minerals are the same as those found in eucrites and diogenites; olivine, pyroxene, and Ca–rich feldspar. These silicate fragments, together with more or less fine-grained metal, formed as breccias. Various levels of recrystallization may result from subsequent events. Mesosiderite textures are therefore highly diverse and can be especially varied for different pieces of a given meteorite.

Fig. 92. Cut and polished surface of the mesosiderite Chinguetti. Metal is bright, silicates dark. About half the natural size. (Photograph Smithsonian Institution)

2.6 The Chemical Significance of Meteorites

In the early 19th century, as the real nature of meteorites was becoming clearer with the passage of time, there was great interest in their chemical characteristics. Chemical analysis, as applied to minerals and rocks, was then a rapidly developing science. The chemical study of these exotic and little understood specimens which fell from the heavens contributed to the understanding of meteorite origins and to the development of both chemistry and mineralogy. These analyses represent the beginning of planetary material science, a science that blossomed again with the return of lunar samples.

In more recent times the field of geochemistry has become of great importance. It is related to chemical analysis, but it is conceptually, a new scientific discipline concerned with the distribution of materials that make up the Earth. Geochemical studies have led to the conclusion that certain groups of meteorites consist of mineral combinations that are similar to our concept of the Earth's interior, a region that will never be accessible to our direct observation. Meteorites present excellent material that may be studied and subjected to experiment on the surface of the Earth, revealing the regularity of material distribution patterns.

The first important result to evolve from these geochemical investigations is that to date no chemical element has been found that is not also present on the Earth, or conversely, the same elements known on Earth are also present in meteorites.

The elements in meteorites can be grouped according to certain, recognized mineral associations in which they occur either exclusively or predominantly. We find one group contained in native nickel–iron, one of the major components of meteorites. Geochemists designate these elements as *siderophile* because of their preference for associating with metallic iron (Greek *sideros* = iron). In addition to iron, these include: nickel, cobalt, copper, gallium, germanium, arsenic, tin, gold; and the platinum metals ruthenium, rhodium, palladium, osmium, iridium, and platinum.

A second group of elements shows a strong relationship with sulfur. Thus, they are concentrated in the iron sulfide or troilite of the meteorites. These elements are called *chalcophile* (Greek *chalkos* = ore). Besides iron (which is both siderophile and chalcophile) the following elements belong to this group: silver, cadmium, indium, thallium, lead, and bismuth; and in addition to sulfur, selenium and tellurium.

A third group shows a strong relationship with oxygen (O), which is particularly concentrated in the silicates of meteorites. We call these elements *lithophile* (Greek *lithos* = stone). Belonging to this group, along with oxygen, are: the alkalies lithium, sodium, potassium, rubidium, and cesium; the alkaline earths beryllium, magnesium, calcium, strontium, barium, and radium; and, additionally, boron, aluminum, scandium, yttrium, and the rare earths; thorium and uranium; and finally silicon, titanium, zirconium, hafnium, vanadium, zinc, niobium, tantalum, phosphorus, chromium, manganese, and a few more rare elements.

A fourth group of elements, those that either alone or in combination are highly volatile, are concentrated in the atmospheres of heavenly bodies and called *atmophile* (Greek *atmos* = air). Examples are

153

hydrogen, nitrogen, and the noble gases, which are found only in very small quantities in meteorites. Nevertheless, some of them, particularly the noble gases helium, neon, argon, krypton, and xenon are of great importance in explaining the early history of meteorites. They will be discussed in Section 3.1.

Just as in the crust of the Earth, there are only a few elements in meteorites that are present in substantial quantities as to be determined by classical analytical techniques. Most elements are present in concentrations of only a few millionths of a gram per gram of meteorite material (1 ppm = 1 part per million = 1 g out of 1 million g, or 1 g/t, corresponding to 0.0001%). Today, there are a number of sensitive methods available that may be used to satisfactorily determine the small amounts of these elements. Spectral analysis has proved to be particularly useful, including X-rays, as well as light in the visible, ultraviolet, and infrared regions. Use is made of the property to emit radiation of characteristic wavelengths under certain stipulated conditions. A simple experiment serves as an example. If one rubs a steel needle across his finger and then holds the needle in the nonluminous region of a gas flame, the flame turns yellow. The reason for this is that a small amount of sodium chloride, which is contained in the perspiration, was picked up by the needle when the skin was rubbed. Yellow is the characteristic color of the element sodium. The rays that the element emits can be directed through a prism and separated into the individual wavelengths that make up the elemental spectrum. The intensities of the characteristic spectral lines may then be used to calculate the concentration. Mass spectrometry uses the ionized atoms of a vaporized sample, which are separated in magnetic and electric fields according to their mass. These atoms can then be counted with electric counters, a very sensitive method to determine trace concentration in small samples.

A particularly powerful new method is that of neutron activation analysis. A sample is placed into a nuclear reactor and irradiated with neutrons, transforming the sought after element (actually only a small fraction of it) into a radioactive atomic species whose radiation can be measured directly. In this manner, many elements can already be measured in the solid sample without interference, others only after dissolving the sample and chemically separating the element of interest. A great advantage of neutron activation analysis is freedom from contamination by impurities that are very difficult to avoid when a sample is chemically processed. When measuring in the ppb range such contaminants can seriously interfere with measurements (1 ppb = part per billion = 1000 million, or 10^{-9} g/g). When the sample is irradiated prior to chemical treatment and only the activated elements are measured, later contamination with nonactive elements presents no problem. It is only since the development of activation analysis that many trace elements have been reliably determined. Uranium, an element that is exceedingly important for age determinations, is an example (see below).

With these previously described methods bulk samples can be analyzed after the meteorite has been ground to a fine powder. The analysis of single mineral grains was formerly only possible after mechanical separation, which was only feasible with grains larger than ca. 0.1 mm. Today, such an analysis is possible with electron and ion microprobes, which employ a finely focused beam of electrons or ions that is directed onto a polished section of a rock. The ion beam sputters a

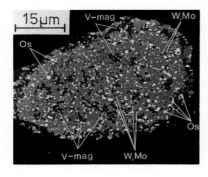

Fig. 93. Metal grain from the carbonaceous chondrite Allende. It contains small inclusions of osmium (*Os*), vanadium-magnetite (*V-mag*), and calcium-tungsten-molybdenum-oxide (*W, Mo*). (After Bischoff and Palme, Geochim. Cosmochim. Acta 51, 1987)

small amount of material into a vapor cloud which can then be analyzed by a mass spectrometer. The electron beam produces X-rays at a point measuring ca. 1 micrometer (1 μm = 1/1000 mm) which are then separated by an X-ray spectrometer into wavelengths characteristic of the elements of interest. X-ray intensitites are related to the amount of element present, making it possible to determine the main elemental composition of a given mineral (to contents of about 0.1%) and to determine the elemental distribution between the various mineral phases. With the ion probe method trace elements in the ppm range and isotope ratios can also be determined. These measurements provide information on the equilibrium states that existed at prevailing temperatures and redox conditions. A simple example is the measurement of the Fe and Mg content of olivine and pyroxene in chondrites that serve to distinguish between H-, L-, and LL-types. Also, the Ni distribution in the taenite of an iron meteorite (Fig. 86) was determined by electron microprobe. A further example is shown in Fig. 93, which is a complex inclusion in the carbonaceous chondrite Allende, presumably originating by condensation in the solar nebula. It was possible to determine its structure and individual phases using electron micro-probe analysis.

2.7 The Cosmic Abundance of the Elements

Table 23 in the Appendix lists the abundance of all the elements in a CI chondrite (Orgueil), an ordinary chondrite (Richardton), an achondrite (the eucrite Juvinas) and, for comparison, the average composition of the Earth's crust. In Table 16 elemental abundances of 15 of these elements are given. Their order of presentation is according to their abundance in the CI chondrite Orgueil, the reason for which will soon become clear. It is noteworthy that O, Fe, and Si in all three meteorites as well as the Earth's crust are by far the most abundant elements, comprising 75 to 80% of the rock (the first ten elements on the list comprise 97 to 99%). There are differences, however, within the listed meteorite types: ordinary chondrites contain less C and S than CI chondrites; the eucrites have still less C and S and very little Ni, but substantially more Ca and Al. How can we derive the *cosmic abundance* from these analyses?

There is another celestial body that needs to be considered in order to establish cosmic abundances: the Sun. It contains 99.9% of all of the material in the Solar System and thus dominates its composition. Employing spectral analysis, the Sun's chemical composition can be determined. It was found that the Sun and CI chondrites have the same relative abundances for all elements (with the exception of the completely or

Table 16. Main elements (wt%) in meteorites and the Earth's crust

Element	Symbol	Orgueil CI	Richardton H5	Juvinas eucrite	Earth's crust
Oxygen	O	44.8	35.1	42.4	47.3
Iron	Fe	18.2	29.0	14.5	3.54
Silicon	Si	10.3	16.3	23.0	30.5
Magnesium	Mg	9.26	13.8	4.0	1.39
Sulfur	S	5.5	1.42	0.20	0.031
Carbon	C	3.4	0.08	0.06	0.032
Nickel	Ni	1.08	1.72	0.0001	0.0044
Calcium	Ca	0.95	1.15	7.7	2.87
Aluminum	Al	0.86	1.05	7.1	7.83
Sodium	Na	0.50	0.71	0.28	2.45
Chromium	Cr	0.26	0.32	0.21	0.007
Manganese	Mn	0.19	0.23	0.40	0.069
Phosphorus	P	0.12	0.10	0.04	0.081
Potassium	K	0.054	0.072	0.022	2.82
Titanium	Ti	0.048	0.06	0.38	0.47

Source: Orgueil from Palme and Beer (1993), Richardton and Juvinas from Palme, Suess and Zeh, Landolt-Börnstein Neue Serie VI/2a 1981, Upper Earth's Crust from Wedepohl, Fortschr Miner 52 (1975).

partially gas-forming elements, such as He, H, O, N, C). This is shown in Fig. 94, where element abundances in the Sun are plotted along one axis and in CI chondrites along the other. All values cluster along a 45° line, where they would be expected to lie if the element abundances are the same for both. As a consequence, the CI values are referred to and used as *cosmic* abundances (more accurately, *solar* abundances) since the CI values can be measured much more accurately than those for the Sun.

Earlier we noted that the carbonaceous chondrites are undifferentiated or primitive meteorites also on the basis of their mineralogical structure. The ordinary chondrites are, in fact, also undifferentiated, but they

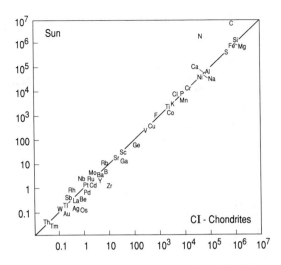

Fig. 94. Abundance of the elements in the Sun versus their abundance in CI chondrites. (After values from Palme, Suess and Zeh, Landolt-Börnstein Neue Serie VI/2a, 1981)

lack some of the lighter, more volatile elements: e.g., C and S and the trace elements germanium, silver, cadmium, indium, thallium, and bismuth. The bulk analysis of chondritic meteorites has shown that in cosmochemistry the *volatility* of the elements is an important factor. *Volatile* elements have low melting and boiling temperatures, the opposite is true for the *refractory* elements. Figure 95 shows the abundance of the elements in different carbonaceous chondrite types arranged according to their volatility. The fact that smooth curves are obatined, and that the chondrite types are distinguished by different ratios of refractory to volatile elements, has led to the idea that condensation from the solar nebula is responsible for these chemical differences. This will be further discussed in Section 3.3.

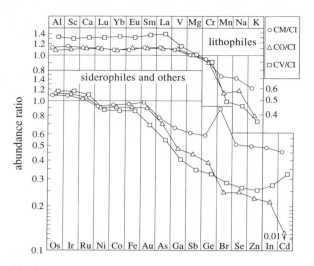

Fig. 95. Mean group atomic abundance ratios normalized to CI and Mg for the CM, CO and CV carbonaceous chondrites. Lithophile elements are plotted in the upper portion, siderophile and other nonlithophile elements in the lower. Elements are ordered from high to low condensation temperatures. Elements left of Mg and Ni-Co-Fe are called refractory, those to the right volatile. (After Kallemeyn and Wasson, Geochim. Cosmochim. Acta 45, 1981)

The differentiated achondrites show stronger deviations from the solar abundance: enrichment in Ca and Al is contrasted with marked depletions in Mg and in the alkalies Na and K; also, the contents in C, S, and Ni are lower. This is analogous to the Earth's crust, which is also enriched in Ca and Al (but also enriched in lithophile elements such as Na and K), and severely depleted in Fe, Mg, and S as well as the siderophile elements (Ni, Co, Ru, Rh, Pd, Os, Ir, Pt). (These severe depletions in siderophile elements, iridium for example, permit the identification of meteorite impact sites on Earth by the iridium enrichment found in impact melt rocks; Sect. 1.6.5.) These geochemical insights and the comparisons with the meteorites have influenced theoretical modeling of the Earth's structure: a metallic nickel–iron core containing the siderophile elements, a mantel of Fe, Mg-silicates, and a crust of the lighter lithophile elements. The Earth as a whole is then assumed to have a chondritic composition.

Despite the differences in the element abundances between the individual meteorite types and the Earth, they demonstrate compliance with the same general laws. A result of modern atomic research is the discovery that we can arrange the chemical elements in a systematic order, each element having its own place and number. Elements are assigned ordinal numbers or atomic numbers. The first element with the atomic number 1 is hydrogen, the one with atomic number 92 is uranium. Atoms with still higher atomic numbers were produced artificially. Every atom consists of a nucleus that contains essentially the atom's complete mass. It is surrounded by a shell of electrons – an electron being the smallest unit of negative electricity – which circle the nucleus in separate orbits. The number of electrons required for an electrically neutral atom depends on the positive charge of the nucleus, which is identical to the atomic number of the element. Hydrogen has 1 electron and uranium 92. The electron shell of an atom is of great importance regarding the chemical behavior of the element and therefore its geochemical characteristics. It has now been shown that the abundance of the elements in the universe is correlated with their atomic numbers. The abundance decreases with increasing atomic number (Fig. 96). The elements hydrogen and helium with atomic numbers 1 and 2, respectively, are most abundant, while uranium (atomic number 92) is a very rare element throughout the universe. There are a few exceptions to this general rule, particularly for the light elements lithium, beryllium, and boron which show an extreme minimum from the general trend of the curve in Fig. 96, and for iron which shows an increase. The conspicuous zigzag form of the curve in Fig. 96 is expressed by the so-called *Harkin's rule*: an element with an even atomic number, for instance 18, 20, 22, is more abundant than its neighbor with an uneven atomic number; 18 is more abundant than 17 and 19, and 20 is more abundant than 19 and 21 (Fig. 96). Both of these principles have been found in the materials of the Earth's crust. The study of meteorites has shown that these are general rules that do not depend specifically on the developmental history of the Earth.

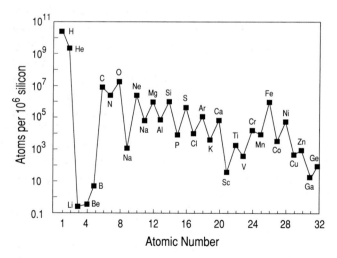

Fig. 96. Solar abundance of the elements, normalized to $Si = 10^6$ atoms. (After values from Palme, Suess and Zeh, Landolt-Börnstein Neue Serie VI/2a, 1981)

A further correspondence between the chemical elements of the Earth and those of meteorites is revealed by their isotope ratios. To understand the concept of isotopes, we must further develop the atomic model presented above. Most elements have several atomic species that are distinguished by their different atomic weights. Atomic nuclei not only contain the positively charged protons, but also small neutral particles of the same weight called neutrons. The weight of an atom is the sum of its protons and neutrons, each of which has a mass of 1. The number of neutrons is approximately equal to the number of protons in a nucleus, but for most elements there are several *isotopes* which have the same number of protons (and therefore the same nuclear charge and atomic number), but a differing number of neutrons. The element magnesium with atomic number 12 has 12 protons, but an individual nucleus may contain 12, 13, or 14 neutrons. These isotopes of magnesium have masses of 24, 25, and 26, designated as ^{24}Mg, ^{25}Mg, and ^{26}Mg. The naturally occuring elements are almost always mixtures of several isotopes, terrestrial magnesium, for instance, consists of 78.6% ^{24}Mg, 10.1% ^{25}Mg, and 11.3% ^{26}Mg. (Only a few elements consist of a single stable isotope; e.g. ^{9}Be, ^{23}Na, and ^{27}Al.) The chemical properties of an element are the same for all of its isotopes, therefore, geochemical processes do not change isotopic ratios. Very precise measurements of both terrestrial samples and meteorites have demonstrated that their elemental isotopic ratios are exactly the same in both types of material.

There are exceptions, however, and these exceptions are of special interest. Even though isotopic contents cannot be modified by *chemical* processes, they can be changed by *physical* processes. Examples are

163

vaporization, condensation, and diffusion, processes that cause only small isotopic fluctuations. Larger effects can occur as a consequence of radioactive processes acting on a specific isotope of an element, and through processes in atomic nuclei that are initiated by highly energetic radiation, such as cosmic rays. These effects within meteorites may be used to determine their ages and their residence time as small bodies in space. This will be discussed in Sections 3.1 and 3.3.

In addition, there are also deviations in isotopic content that cannot be explained as a consequence of radioactive or nuclear-physical processes. These so-called *isotope anomalies* are considered evidence for material from outside our Solar System, material that originated from different nuclear processes in stars and was not mixed homogeneously with the solar nebula as its first solid components formed.

2.8 Isotope Anomalies

The most striking anomaly occurs in the most abundant element in both meteorites and the Earth, oxygen. It consists of the three isotopes ^{16}O (99.76%), ^{17}O (0.038%), and ^{18}O (0.20%). These average isotope contents vary in terrestrial samples up to 10%, caused by the previously mentioned physical processes of condensation, vaporization, and diffusion, or separation by different bonding forces to other elements in certain minerals. These fractionations are proportional to the mass differences between the isotopes, being half as strong for ^{17}O as for ^{18}O when compared to ^{16}O. Presenting the deviation of the

$^{17}O/^{16}O$ content of terrestrial rocks (compared to a standard value and presented as $\delta\ ^{17}O$ in per mill) against the deviation of the $^{18}O/^{16}O$ content ($\delta\ ^{18}O$), the values lie on a straight line with a slope of $1/2$ (Fig. 97). Meteorite values must lie along this same line if they derived from the same original oxygen reservoir by the physical processes mentioned above. It has now been demonstrated that the various meteorite classes assume their own positions on the diagram, indicating that each of them must have derived from separate reservoirs with their own individual initial isotope ratios. In contrast, samples of the Moon lie along the Earth line, an argument for the same origin of the Earth and the Moon. Very near this line, but clearly separated from it, is the region for eucrites, diogenites, and howardites as well as the SNC meteorites. The greatest deviation is shown by the carbonaceous chondrites. Their high temperature minerals (olivine, pyroxene, spinel, melilite) do not lie along a fractionation line with slope $1/2$, but along a line with a slope of 1. This is probably a mixing line with a pure ^{16}O-component that these minerals contain in different amounts.

Other isotope anomalies have been found in certain components of the carbonaceous chondrites, most importantly in Ca, Al-rich inclusions. These involve the elements Si, Mg, Ti, Ca, and others. These anomalies probably originated from material introduced from outside our Solar System. The American astrophysist D.D. Clayton named it *star dust*, which was apparently introduced by a Supernova very early in the history of the Solar System. This star dust evidently contained diamonds and silicon carbide. These minerals have been found as minute grains of about 50 Å in the matrix of carbonaceous chondrites. Their interstellar origin is

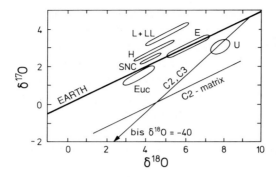

Fig. 97. Isotope ratios of oxygen for different meteorite classes and the Earth, after measurements of R.N. Clayton and coworkers from the University of Chicago. $\delta\ ^{17}O$ and $\delta\ ^{18}O$ are the deviations of the $^{17}O/^{16}O$ and $^{18}O/^{16}O$ ratio, respectively, from the standard value of mean ocean water (SMOW) in ‰. The ordinary chondrite classes are marked by the abbreviations *H*, *L*, and *LL*; *E* enstatite chondrites; *C2* and *C3* carbonaceous chondrites; *Euc* eucrites, howardites and diogenites; *SNC* shergottites, nakhlites and chassignites; *U* ureilites

also confirmed by further isotope anomalies, particularly in the rare gas xenon that is contained within them.

2.9 Organic Compounds

The primitive carbonaceous chondrites, particularly, types CI and CM, contain organic substances. They are compounds of carbon with hydrogen, oxygen, nitrogen, and sulfur (Fig. 96 shows that these are among the most abundant elements in the cosmos), just as they are found in living organisms. It was believed earlier that they could only be derived from living matter, but later it was found that they can also be produced in the laboratory from the elements or inorganic compounds. Therefore, the presence alone of organic material in meteorites is not proof of life on meteorite parent bodies. Nevertheless, this question must be examined in more detail in order to determine whether or not organic life existed.

The organic material in carbonaceous chondrites consists to a major extent of low soluble polymeric material (kerogen), but easily soluble compounds including, among others, amino acids, have also been found (Table 17). Because of their biological importance, the identification of amino acids led to a vigorous scientific debate: did they originate from organisms or could they have been produced inorganically? Here, the *optical activity* of these amino acids plays an important role, as organisms can produce optically active substances that in solution rotate the polarization plane of a beam of polarized light as it passes through. The same compounds inorganically produced do not display this property.

Table 17. Carbon compounds in the CM chondrite Murchison

Acid-insoluble polymer	1.45%
Carbonates	0.1–0.2%
Hydrocarbons	30–60 ppm
Monocarboxylic acids	330 ppm
Amino acids	10–22 ppm
Primary alcohols	11 ppm
Aldehydes	11 ppm
Ketones	16 ppm
Amines	11 ppm
Urea	25 ppm
Purines	1.2 ppm

Source: Mullie and Reisse, Topics Curr Chem 139 (1987).

The optical activity in these compounds is caused by their molecular structure. A molecule can assume two mirror image forms that have the same relationship to each other as a left-hand glove to the right-hand one. In inorganically produced substances there are always equal amounts of the right-hand and the left-hand form (resulting in optical inactivity), while in organic synthesis one form is preferred over the other.

Early measurements appeared to indicate the presence of optical activity in amino acids from carbonaceous chondrites, but later work established its absence. Furthermore, it can be shown that most of the organic compounds in carbonaceous chondrites could have been synthesized by catalytic reactions of carbon monoxide and hydrogen on water-containing silicates or magnetite (which are present in carbonaceous chondrites) under conditions existing in the solar nebula. They were almost certainly not formed from living organisms. To the contrary, the inverse is much more probable: i.e., these organic compounds from meteorites may have provided the raw materials for the development of life on Earth (and perhaps on other celestial bodies). Carbonaceous chondrite material was certainly incorporated into the Earth and the other planets as they formed, so that an early, primitive ocean already had the building blocks of life at its disposal.

3 The Origin and Formation of Meteorites

3.1 Meteorite Ages

The materials from which meteorites were formed have had a complex history, which is presented schematically in Fig. 98. The beginning is *nucleosynthesis*, i.e., the production of the chemical elements by nuclear processes under conditions of very high temperatures and pressure in the interior of stars. All of the elements originated from hydrogen through the addition of protons, neutrons, and electrons. Over time, these materials were distributed throughout interstellar space. A localized cloud of higher density could have formed and through a process referred to as *gravitational collapse* produced the *solar nebula*. With further contraction and increase in density, the nebula evolved into the Sun and numerous small solid bodies (planetesimals). In a final stage, many of the planetesimals combined to form a smaller number of larger bodies, the planets. In such a body, referred to as the *meteorite parent body*, the history of a meteorite as an isolated piece of rock actually begins. Later, separated from its parent body, presumably by an impact, the meteoroid began its life as a small body in its own orbit around

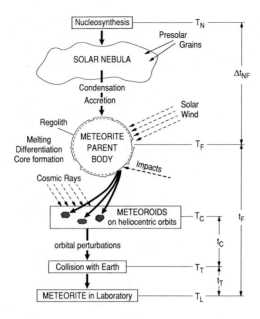

Fig. 98. Schematic history of meteoritic matter with time scale. (After Voshage, J. Mass Spectrometry 1, 1968)

the Sun. This orbit eventually evolved in such a manner that the gravitational forces of the Earth caused the meteoroid to plunge to the ground as a meteorite.

One can attempt to date the various stages in the history of a meteorite and to determine appropriate "ages". In the following we will focus on the history of the meteorite after the formation of the parent body. We can distinguish the following ages (see Fig. 98), which can be determined by the exact analysis of the meteorite's trace elements and their isotopic ratios:

Fig. 99. Apparatus for the determination of very small quantities of helium and argon designed by F.A. Paneth. (After Paneth, Endeavor, 1953)

1. The formation age (t_F) of the meteorite as a rock,
2. The cosmic ray exposure age (t_C), or how long the meteorite has spent in its orbit in space as a meter- to centimeter-sized body,
3. The terrestrial age (t_T), i.e., the time from its fall to Earth to today.

The noble gases are particularly useful for age determinations. One reason is that for every period of meteorite history there is a characteristic noble gas isotope. Furthermore, noble gases do not take part in geochemical fractionations, because they do not react with other chemical elements (hence *noble*). Thirdly, they are easily extracted from a sample by heating and separated from other elements, so that they can be readily measured in very small amounts. This property also resulted in their escape from solid materials when they formed hot, so that practically all of the noble gases found in solid samples are newly formed. Examples will be discussed below.

A pioneer in the measurement of the noble gases was the chemist, F.A. Paneth (1887 to 1958), who developed very sensitive methods for their separation and measurement during the years 1930 to 1950. With the apparatus shown in Fig. 99 in his laboratory at Durham University, England, and later the Max-Planck-Institute for Chemistry in Mainz, Germany, he could measure helium or neon at levels of $10^{-7}\,cm^3$ with an accuracy of 1 %. With today's mass spectrometers, amounts 100 times smaller may be determined.

3.1.1 Formation Age

The formation age gives the most important date in the history of a meteorite. It indicates the point in time when the meteorite completely crystallized (achondrites, iron meteorites) or at which its components formed (chondrules, nickel–iron, or matrix in the case of chondrites). We can measure the formation age with the help of the natural radioactive decay of certain elements. One method uses the decay of uranium and thorium to lead (Pb) and helium (He). Three decay series are involved here:

$$^{238}\text{Uranium} \rightarrow\ ^{206}\text{Pb} + 8 \text{ atoms of } ^{4}\text{He}$$
$$^{235}\text{Uranium} \rightarrow\ ^{207}\text{Pb} + 7 \text{ atoms of } ^{4}\text{He}$$
$$^{232}\text{Thorium} \rightarrow\ ^{208}\text{Pb} + 6 \text{ atoms of } ^{4}\text{He}$$

The disintegration rate is constant; it is given as the *half-life*, the time in which half of the atoms of a certain element decay. For ^{238}uranium this is 4.51

billion years. As all rocks contain small amounts of uranium and thorium, their helium content increases slowly over time. Measurements of He, uranium, and thorium content permit this time period to be calculated as an age. It represents the length of time during which the fission product helium cannot escape from the rock; this is generally the time period since the rock was formed by solidification from a melt.

Naturally, one may also determine ages from the quantities of the lead atoms produced, ^{206}Pb, ^{207}Pb, and ^{208}Pb. Age determinations using this *lead method* are rather difficult, however, because the quantities of radiogenic lead are very small and must be measured in ultra-clean laboratories to prevent contamination with common lead. Careful measurements give precise results, however. The helium method is simpler but it also has a serious disadvantage: helium can easily be lost from a sample by diffusion, particularly by a later heating or shock event. This is also true of the potassium/argon method that relies on the decay of ^{40}K to the noble gas ^{40}argon. Thus, reliable formation ages can be obtained by these methods only for undisturbed samples. Less sensitive to such errors is a fourth method which is based on the decay of ^{87}rubidium to ^{87}strontium. In Table 18 formation ages are presented using different methods on different meteorite types.

The results show that chondrites, achondrites, and iron meteorites all have similar ages of around 4500 million years (with the important exception of the achondrite Shergotty, which will be discussed later). This same age has been found for the Earth by lead isotope measurements and is therefore considered the age of the Solar System. This old age, however, does not apply to individual rocks of the Earth, but only to the Earth as a whole. The rocks of the Earth's crust have

Table 18. Formation ages of meteorites, in million years

Meteorite	Class	Method	Age		Author
Guarena	H5-chondrite	Rb/Sr	4460	± 80	Wasserburg et al. (1969)
St. Severin	LL6-chondrite	Sm/Nd	4550	± 330	Jacobsen and Wasserburg (1984)
		Pb/Pb	4553.6	± 0.7	Göpel et al. (1991)
Tennasilm	L4-chondrite	Pb/Pb	4552	± 13	Unruh (1982)
Bruderheim	L6-chondrite	Pb/Pb	4535	± 4	Unruh (1982)
Indarch	E-chondrite	Rb/Sr	4460	± 80	Gopalan and Wetherill (1970)
Juvinas	Eucrite	Rb/Sr	4500	± 70	Allègre et al. (1975)
Moama	Eucrite	Sm/Nd	4460	± 30	Jacobsen and Wasserburg (1984)
Ibitira	Eucrite	Rb/Sr	4420	± 250	Birck and Allègre (1978)
		Pb/Pb	4556	± 6	Chen and Wasserburg (1985)
Serra de Magé	Eucrite	Sm/Nd	4410	± 20	Lugmair et al. (1977)
Kapoeta	Howardite	Rb/Sr	4440	± 120	Papanastassiou and Wasserburg (1976)
Norton County	Aubrite	Rb/Sr	4390	± 40	Minster and Allègre (1976)
Shergotty	Shergottite	Rb/Sr	360	± 16	Jagoutz and Wänke (1986)
Estherville	Mesosiderite	Rb/Sr	4542	± 203	Brouxel and Tatsumoto (1991)
		Pb/Pb	4555	± 35	
		U/Pb	4560	± 31	
		Sm/Nd	4533	± 94	
Vaca Muerta	Mesosiderite	Pb/Pb	4536	± 15	Ireland and Wlotzka (1992)
Colomera	Octahedrite	Rb/Sr	4510	± 40	Sanz et al. (1975)

formation ages of a few million years and very few are older than 2000 million years.

The great age of the meteorites means that the rock formation process of the parent body must have been completed very soon after the formation of the planetary system. This is only possible for relatively small bodies, i.e., those that could cool quickly enough. This will be discussed further in Section 3.2.

3.1.2 Cosmic Ray Exposure Ages

During their flight through space as small bodies, meteorites are exposed to cosmic rays. This radiation consists essentially of energetic protons that produce nuclear reactions in meteorites. These are the so-called spallation or nuclear splitting reactions whereby atoms are split into new, lighter nuclei. The number of transitions is actually very small, so that after millions of years only very few of these new atoms are detectable. The cosmic rays penetrate only about 1 m into the solid material and do not penetrate the Earth's atmosphere. As a consequence, the reaction is "turned on" only after the meteorite becomes separated from its parent body (through the impact of another body) and travels alone as a small body in its orbit around the Sun. The effect is then "turned off" when it lands on the Earth. The longer the meteorite is exposed to cosmic rays, the greater the production of spallogenic atoms. The determination of cosmic ray ages depends again on measurements of noble gases, especially ^{3}He, ^{21}Ne, and ^{38}Ar. The production rate of these isotopes, as a result of energetic proton reactions, has been determined through experiments.

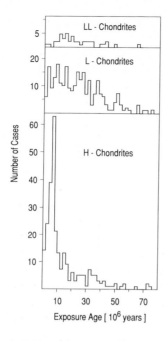

Fig. 100. Cosmic ray exposure ages for LL-, L- and H-chondrites. (After L. Schultz, 1987)

Figure 100 presents cosmic ray exposure ages determined in this manner for the three types of ordinary chondrites. Their ages range approximately between 1 and 60 million years. The L-chondrites show a wide distribution with most values between 10 and 40 million years, while a clear maximum at 5 million years is seen for the H-chondrites. One must therefore conclude that the H-chondrite parent body experienced a large collision 5 million years ago, resulting in the production of many meteorites. Apparently, the L-parent body experienced a continual series of many impacts.

The measurement of cosmic ray exposure ages in iron meteorites also depends on the noble gases as well as other elements, for instance, the two

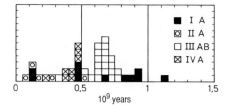

Fig. 101. Cosmic ray exposure ages for iron meteorites of different chemical groups. (After Voshage et al., Z. Naturforsch. 38a, 1983)

potassium isotopes ^{41}K (stable) and ^{40}K (radioactive). The amount of natural potassium in the iron is so small that the effects of cosmic ray interactions become measurable. Figure 101 shows the distribution of cosmic ray exposure ages of iron meteorites determined by the potassium method. They are about ten times as high as those for the stone meteorites, namely, 100 million years to 1000 million years. Several classes again show age clusters, the medium octahedrites (chemical group IIIAB) at 650 million years, the fine octahedrites (chemical group IVA) at 400 million years. These groupings show that all the meteorites of this type originated from the same parent body and were separated from it in the same event. The great difference in cosmic ray exposure ages between the irons and the stones is probably related to the fact that stones undergo more rapid destruction by erosion and impacts in space than the more resistant irons.

The various long-lived radioactive isotopes that are produced in meteorites by cosmic rays may also provide evidence regarding the temporal constancy of this radiation through the history of the Solar System. A long-lived isotope such as ^{40}K, HL (= half-life) 1.3×10^9 years, integrates the effects of cosmic rays over a substantially longer time than would ^{36}Cl, HL 300000 years, or ^{39}Ar, HL 270 years. From such comparisons it has been shown that the intensity of the cosmic rays in the Solar System has been constant over the last 1000 to 2000 million years, but that in about the last 10 million years it has been around 50 % higher.

3.1.3 Terrestrial Ages

The presence of cosmic ray-induced radioactive isotopes presents the opportunity to measure terrestrial ages, i.e., how long the meteorite has lain on the ground after its fall. At the time of fall of the meteorite the radioactive isotopes produced by spallation reactions have normally reached a *saturation value*, i.e., atoms are decaying and forming in equal numbers. On the ground, the cosmic rays are "turned off" by atmospheric shielding and the radioactive atoms decay with their characteristic half-life (HL). Useful isotopes include ^{39}Ar (HL 270 years), ^{14}C (HL 5700 years), and ^{36}Cl (HL 300000 years). Therefore, if the content of such an isotope is measured in a meteorite find and compared with the saturation value of freshly fallen meteorites of the same type, the time since its fall can be calculated.

Table 19 lists several terrestrial ages for stone and iron meteorites. It shows that stony meteorites can reach lifetimes of up to 35000 years in desert areas. Tamarugal, found in 1903, is the oldest iron meteorite dated so far; it was exposed to the extremely dry climate of the Atacama Desert in Chile for 1.5 million years. Through measurement of the short-lived ^{39}Ar it was possible to prove that the chondrite Benthullen, which was found in 1951 near Oldenburg, Germany,

Table 19. Terrestrial ages of meteorites

Meteorites		Class	Isotope	Age, years
Irons:	Keen Mountain, Virginia, USA	Hexahedrite	^{39}Cl	1300
	Clark County, Kentucky, USA	Octahedrite Om	^{36}Cl	600000
	Tamarugal, Chile	Octahedrite Om	^{36}Cl	> 1.5 million
Stones:	Daraj 008, Libya	H-chondrite	^{14}C	3500
	Eunice, Texas, USA	H-chondrite	^{14}C	8000
	Rock Creek, Texas, USA	L-chondrite	^{14}C	15500
	Daraj 114, Libya	H-chondrite	^{14}C	16300
	Finney, Texas, USA	L-chondrite	^{14}C	27000
	Daraj 119, Libya	L-chondrite	^{14}C	35000

Source: Irons: Vilcsek and Wänke, IAEA Vienna (1963). Stones: Jull et al., Geochim Cosmochim Acta 54 (1990) and unpublished.

fell less than 200 years ago and cannot be the meteorite reported to have fallen there in 1368. In general, however, the determination of terrestrial ages is not precise enough to identify meteorite finds with historical fireball reports.

3.2 The Origin of Meteorites

What conclusions can now be drawn concerning the genesis and origin of meteorites from the mass of carefully collected and synthesized material available? This is the last and most important question that will be considered.

As we have seen in the historical overview, during the time of Chladni, the idea that meteorites originated on Earth, perhaps by ejection from volcanoes, was rejected. Our modern views of the composition and structure of rocks that can be found on the surface of the Earth preclude a terrestrial origin of meteorites. Only for specific types of achondrites could this be considered a possibility, but these show such a close relationship to the rest of the meteorites that a special origin for them is improbable. Meteorites are undoubtedly extraterrestrial bodies. Today, we can also prove by the content of cosmic ray-produced isotopes that these rocks have traveled through space.

Far more difficult for us to determine is whether they belong to our Solar System or whether they come from interstellar space, a question that is of great importance for our conception of the uniformity of material in the universe.

First, what is the relationship between shooting stars and fireballs, the celestial bodies that do not reach the surface of the Earth and from which we receive only light signals? All observations indicate that shooting stars, fireballs, and meteorites represent essentially the same phenomenon. As mentioned in Section 1.7.3, no relationship between meteorites and the so-called meteor showers could be proven. There are enough sporadic shooting stars where the intensity of the emitted light gradually merges into the range of fireball intensities. These in turn show all of the properties of those fireballs that produce meteorites which fall to Earth. Between the two types of occurrences, it appears that the only real difference is that of the mass of the object: an extraterrestrial body can only survive the difficult process of penetrating the atmosphere when it has a specific mass, if too small, it burns to dust and smoke.

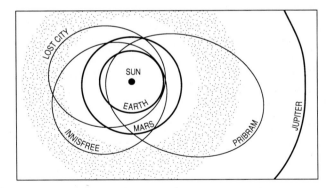

Fig. 102. Orbits of the three meteorites, Pribram, Lost City, and Innisfree, which were determined from their photographically recorded paths through the terrestrial atmosphere. All three originate in the asteroid belt (*dotted*)

Earlier, orbital determinations for shooting stars, fireballs, and meteorites available to astronomers relied only on eyewitnesses reports and were therefore highly imprecise. The photographic recordings of the three fireballs from Pribram, Lost City, and Innisfree have permitted a fundamentally better reconstruction of the orbits for these meteorites. As shown in Fig. 102, all three originated in the asteroid belt.

3.2.1 Asteroids

Asteroids are small planets with diameters of several to a few hundred kilometers, which populate a wide belt between Mars and Jupiter. Ceres (discovered in 1801) was the first and largest asteroid to be discovered (diameter 1000 km). Several years earlier astronomers noted that there should have been a planet between Mars and Jupiter. The newly discovered Ceres was at first seen as the missing planet, but then over the following 6 years three more planets were discovered: Pallas, Juno, and Vesta with diameters from 250 to 600 km. At that time it was believed that these bodies were the debris of a former larger planet. After 1845 ever-increasing numbers of asteroids were found; up to 1900 425 were known, and today more than 5000 are known. Included are bodies with diameters of only 2 km, e.g., Icarus, discovered in 1949.

The spectra of reflected sunlight from individual asteroids indicate that they are of various types. These spectra give information on the composition of the surfaces of the asteroids. The S-type probably consists of silicates and nickel-iron and could be the parent body of stony iron meteorites and also chondrites.

The C-type is very dark and could have produced the carbonaceous chondrites. About ten distinct types are recognized that probably all have different compositions. Consequently, the concept that they are debris from a single planet is replaced with the idea that asteroids are independently formed bodies. Included among them could also be outgassed comet nuclei that can no longer produce visible trails. It is known that comets are rich in volatile elements and water. This has given rise to speculation that the water-containing carbonaceous chondrites of types C1 and C2 may originate from such comet nuclei.

Through the effect of gravity the planets, particularly Jupiter, can disturb the orbits of asteroids so that they cross the orbit of Mars and even that of the Earth. Collisions between and impacts upon asteroids produce ejected pieces that in the same manner could achieve orbits that cross that of the Earth. It must also be assumed that every meteorite class originates from its own parent body. Only for some types are associations sufficiently close to make an origin from the same parent body probable. Thus eucrites, howardites, and diogenites probably come from the same asteroid, most likely Vesta. It is possible that the mesosiderites also come from this body. Another association is considered for the SNC meteorites, all of which must have originated from a large body, perhaps Mars (see below). Relationships exist also between the carbonaceous chondrites and the ureilites, on the one hand, and between the pallasites and the iron meteorites of groups IIA and IIB, on the other. The stone meteorites apparently account for 12 different parent bodies, while 8 of the chemical groups of iron meteorites require their own parent bodies. In addition, one must assume that each of the approximately 50 "anomalous" iron meteorites originate from a separate parent body. Thus, a minimum of 20 and possibly as many as 70 different bodies in our Solar System exist from which we now have sample material on Earth thanks to meteorite falls.

Asteroids with diameters of a few hundred kilometers are small compared to the Earth (diameter 12 750 km) and to the Moon (diameter 3476 km). There are other indications that most meteorites originate from such small bodies. One is the absence of high pressure minerals known from the deeper strata of the Earth's crust. Diamonds, which are formed on the Earth by high pressure, are formed in meteorites by shock waves acting on graphite. Another is the cooling rates of iron meteorites and chondrites (1 to 100°C/million years) which are only possible on bodies of asteroidal size. This is also supported by the old ages of meteorites that date the time of cooling to a temperature at which the escape of helium or of an isotopic exchange through diffusion in the minerals is no longer possible. Large bodies, on the other hand, such as the planets, have not cooled sufficiently since their formation, and are thus still active, magmatic planets. The volcanoes of the Earth are proof of this.

We have seen that the majority of meteorites come from the asteroids (a few also originate from Mars and Moon, see below), i.e., they originate within our Solar System. This is also shown by the identical age of 4500 million years for the Earth and the meteorites, which is also the age of the Solar System. A further argument is the uniformity of the isotopic ratios found in meteorites, the Earth, and the Moon. The interstellar grains with deviating isotopic contents in carbonaceous chondrites are only minor admixtures.

Isotopic anomalies in the major elements such as oxygen are accounted for by inhomogeneities in the solar nebula.

3.2.2 Meteorites from Mars and the Moon

The SNC meteorites were mentioned earlier in Section 2.5.1.2.2. These achondrites are younger than 1.5 billion years, distinctly younger than the chondrites and the other groups of achondrites. The best studied is Shergotty (see Fig. 78), for which a formation age of 360 million years has been measured. It must therefore have originated from a large body that has remained magmatically active until comparatively recently. This naturally led to the consideration of Mars. Recently, a direct comparison with Mars became possible by employing determinations of the noble gases: glass fragments in an Antarctic shergottite contain rare gases and nitrogen in the same amounts and isotope ratios as those measured in the atmosphere of Mars by the Viking Mars probe. Also, the analytical chemical data, which were transmitted from the probe to Earth, compared very well with the analysis of Shergotty (when only the silicate portion is taken into account). Thus, the detailed study of the shergottites and the other SNC meteorites can provide an important reference point regarding the composition and structure of the planet Mars.

In order to be ejected from Mars, a piece of rock must be accelerated by an impact event to the *escape velocity*, which is 5 km/s. As a consequence, extreme shock forces are experienced, leading to the belief that the rock must be melted or pulverized. It was

Fig. 103. The lunar meteorite ALHA 81005, weight 31.4 g, found in Antarctica, scale cube is 1 cm. (NASA photograph)

therefore considered impossible that ejecta from Mars could reach the Earth intact. The shergottites show strong shock effects (feldspar has been transformed to glass), but they have not been completely melted. For meteorites from the Moon the same considerations would apply, except that here the escape velocity is only half as large (2.4 km/s). The 31.4-g stone ALHA81005 found in 1981 in Antarctica is similar in appearance to a brecciated achondrite (Fig. 103), but based on its composition and mineralogy, it is undoubtedly a sample from the lunar highlands. Since then, more lunar meteorites have been found in Antarctica and one also in Australia (Calcalong Creek). Meteorites from the Moon have landed on the Earth, therefore, it seems probable that also pieces from Mars could have reached the Earth's surface.

3.3 The Formation of Chondrites

We have already seen that a solar nebula formed after the process of nucleosynthesis, which then evolved into our planetary system. The time span between the end of nucleosynthesis and the formation of solid bodies is called the *formation interval* Δt_{NF} (Fig. 98). This interval gives another *age* that can be measured with radioactive isotopes by employing the iodine-xenon system.

3.3.1 Iodine-Xenon Ages

Short-lived radioactive nuclei were formed during nucleosynthesis; they decayed rapidly into other atomic species and are now extinct. If they were incorporated into solid material while they were still alive, their fission products must have accumulated in that material and be detectable. This is the case for radioactive iodine, ^{129}I, which has an HL of 16 million years and decays to the stable isotope ^{129}Xe, a noble gas. The ^{129}I was incorporated together with normal iodine into certain minerals and later decayed to ^{129}Xe, yielding minerals containing both normal iodine and ^{129}Xe. Through the stepwise heating of a sample containing iodine and the analysis of the released iodine and xenon isotopes, it can be demonstrated that iodine and ^{129}Xe are correlated to each other. In order to use the measured ^{129}Xe to calculate the time between nucleosynthesis and the beginning of xenon retention, the initial ratio of normal ^{128}I to ^{129}I must be known. This ratio can only be estimated from model calculations for nucleosynthesis, leading only to an approximate determination of the time period. Such calculations for the chondrite Bjurböle yield the relatively short formation interval of about 60 to 200 million years. Determinations of the relative differences in iodine-xenon ages between various meteorites are more precise. Such calculations suggest that the chondrites of various types formed within a period of only 25 million years.

3.3.2 Condensation and Ca,Al-Rich Inclusions

The temperature in the solar nebula was high enough that all of the solid material vaporized with the exception of a small number of interstellar grains that are recognized in chondrites. Solid materials formed by condensation during slow cooling. The condensation sequence followed the boiling points of the elements and their compounds. Table 20 lists the chemical compounds in the sequence in which they would condense from a gas of solar composition. A pressure of 10^{-3} atm is assumed, a lower pressure would also lower the condensation temperature.

The first condensates are calcium- and aluminum-rich oxides and silicates. In the descriptions of carbonaceous chondrites we have already seen that perovskite, melilite, and spinel are present in the Ca,Al-rich inclusions. These observations led to the assumption that these inclusions are the first condensation products of the solar nebula. The enrichment of refractory elements (i.e., with high melting and boiling points) extends also to trace elements such as the rare earths and the siderophile elements. Figure 104 illustrates the systematic enrichment by a

Table 20. Condensation temperatures in the solar nebula

Mineral	Composition	Temperature (°C)
Corundum	Al_2O_3	1485
Perovskite	$CaTiO_3$	1374
Melilite	$Ca_2Al_2SiO_7 - Ca_2MgSi_2O_7$	1350
Spinel	$MgAl_2O_4$	1240
Metal	Nickeliron	1200
Diopside	$CaMgSi_2O_6$	1180
Forsterite	Mg_2SiO_4	1170
Anorthite	$CaAl_2Si_2O_8$	1090
Enstatite	$MgSiO_3$	1080
Troilite	FeS	430
Magnetite	Fe_3O_4	130

Source: L. Grossman, Geochim Cosmochim Acta 36 (1972).

factor of 20 for various refractory elements in an inclusion from the Allende meteorite. Refractory metals such as iridium, osmium, and platinum form small metal grains during condensation that become embedded in the silicates. An example of such a complex metal grain is illustrated in Fig. 93. Figure 105 shows the condensation curves for the noble metals. Osmium, tungsten and rhenium condense first and form an alloy that upon further reduction in temperature incorporates molybdenum, iridium, ruthenium, and platinum. Only later, at lower temperatures, are the much more abundant elements nickel and iron included.

According to Table 20, magnesium silicates such as forsterite and enstatite should condense after the Ca, Al-rich phases. They are prominent in the chondrules in chondrites. Do these chondrules also form by condensation from the solar nebula?

189

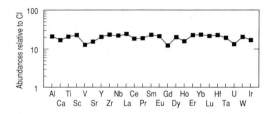

Fig. 104. Enrichment of refractory elements by a factor of 20 in a Ca, Al-rich inclusion from Allende. (After Wänke et al., Earth Planet. Sci. Lett. 23, 1974)

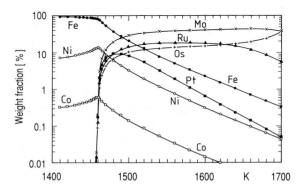

Fig. 105. Condensation curves for refractory metals in the solar nebula. These metals form alloys; with falling temperature they become enriched in Fe, Ni, and Co, which dominate below 1460 K. (After Palme and Wlotzka, Earth Planet. Sci.

3.3.3 The Origin of Chondrules

Despite extensive research the origin of chondrules is still not understood. They are rapidly solidified melt droplets. But where and how did they originate? There are essentially two theories. The first theory postulates that chondrules originated as primary products of the solar nebula, either by direct condensation or by melting of condensed dust. The second proposes the formation of chondrules on a parent body, where they are formed from dust or coarse-grained rock secondarily through impacts. However, neither theory can explain all of the properties of chondrules.

Figure 106 shows the contents of selected elements in chondrules from the carbonaceous chondrite Allende. It is clearly seen that major elements like Mg (range 8–26%), Al (range 1–13.5%), Ca (range 1–10%), and Fe (range 2–20%), as well as trace elements like As (range 0.2–1.5 ppm), vary over a wide range. This chemical variation already poses the first problem for the condensation mechanism: direct condensation from a homogeneous gas should yield uniform compositions. This problem can be overcome, if one assumes that at first different, rather coarse-grained minerals condensed. These aggregated in various proportions into dust balls. Transient, perhaps only local heating events (lightning discharges, frictional heating) then melted these inhomogeneous aggregates, thus forming chondrules. The second problem is the presence of iron oxide in the chondrule silicates. Through direct condensation from the solar nebula this is not possible. Because of insufficient oxygen, iron condenses as metal (see Table 20), and the condensing olivines and pyroxenes contain only MgO. The chondrules contain also volatile elements such as Na, K, and e.g. arsenic (see Fig. 106) that should not condense together with the magnesium silicates. One can overcome these difficulties employing the hypothesis that the oxygen content of the solar nebula was increased by a special process and that the volatile elements were introduced into the chondrules later (perhaps on the parent body). A primary formation of the chondrules would naturally

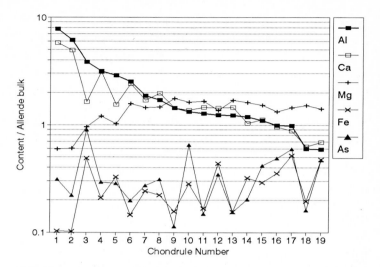

Fig. 106. Composition of chondrules from the CV3 chondrite Allende, plotted relative to Allende bulk. The chondrules are arranged with falling Al content. Ca follows Al, whereas Mg increases with falling Al. Fe and As are depleted in the chondrules compared to the bulk, but not absent. (After analyses by Palme and Spettel, MPI Chemie, Mainz)

fit well with the primitive, undifferentiated character of the chondrites.

A secondary origin for the chondrules on the surface of a parent body by the impact melting of rocks consisting of olivine, pyroxene, and some feldspar would easily explain the chemical compositions shown in Fig. 106. In addition, chondrule-like spherules are known from the Moon's regolith and terrestrial impact rocks, although only in very small quantities. The parent material should then be chemically chondritic but without chondrules. Such a rock has never been found among the meteorites, except perhaps the chondrule-free and still primitive C1 chondrites. Detailed studies of the trace element distribution in the minerals of chondrules by ion probe have shown, however, that their formation from preexisting minerals from an igneous rock on a parent body is rather unlikely. Such minerals should have a "geochemical" trace element distribution pattern; for example, scandium should be preferentially found in pyroxene, and pyroxene-rich chondrules should also be scandium-rich. This is not the case.

3.3.4 Matrix

Finally at very low temperatures in the range of 500°C, sulfide, magnetite, water-containing silicates, and organic compounds condense as fine dust, along with the volatile elements, in particular indium, bismuth, and thallium. In this way the matrix of the carbonaceous chondrites could have originated, containing all of these constituents. The ordinary chondrites also have a small amount of this matrix. Therefore, two components are found in the chondrites, a high temperature component that consists essentially of chondrules (including a small fraction of Ca,Al-rich inclusions), and a low temperature component, the matrix with the volatile elements. The depletion of the ordinary chondrites in volatile elements, water, and carbon would then result from the presence of only a small amount of matrix. It is also possible that these chondrites originally contained the same amounts of volatile elements but lost them later due to a heating event.

3.4 Planetesimals and the Formation of Achondrites

From the material resulting from the condensation process, e.g., fine-grained dust and possibly also chondrules, small bodies formed by agglomeration. These so-called planetesimals had diameters from a few to tens of kilometers. It is plausible that with increasing distance from the Sun the condensed materials became more oxidized. Chondrites then probably formed in the inner Solar System, while the more oxidized carbonaceous chondrites formed further out where the temperature was also lower. Some of these small bodies remained undifferentiated, because they never became hot enough to melt. They thus formed the parent bodies of the chondrites. The temperatures that these chondrites experienced can be measured by geological thermometry techniques. One such approach is shown in Fig. 107. It employs the distribution of iron and magnesium between olivine and chromium spinel, which is dependent upon temperature. For the partly equilibrated chondrites of type 3 the value is 600 to 700°C, and for the equilibrated types 4 and 5, 800°C. The carbonaceous chondrites, however, experienced much lower temperatures. Their contents of volatile elements, water-containing silicates, and easily decomposed organic compounds make it clear that they could not have experienced temperatures above ca. 500°C since their agglomeration. It is this stroke of good fortune that permits us to study even today the wide variety of unchanged materials from the early period of the Solar System.

Other bodies were heated to such high temperatures that they started to melt. Due to its higher

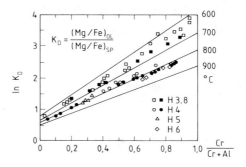

Fig. 107. The distribution of Mg and Fe between olivine (*OL*) and spinel (*SP*) is plotted against the Cr/(Cr+Al) ratio in spinel for different types of chondrites (*H3, H4, H5, H6*). The data points form isotherms showing the equilibration temperature of these chondrites. (After Wlotzka, Meteoritics 22, 1987)

density a metal and sulfide melt could form pools which sink to form ultimately a nickel-iron core. Silicate melts formed a mantle of magmatic rocks that could then become differentiatied through the sinking of heavy crystals or by partial melting. Iron meteorites, stony iron meteorites, and achondrites may all have originated from such a body.

The heat source was probably provided by short-lived radioactive elements, for example, the isotope ^{26}Al that with a half-life of 720 000 years decays to ^{26}Mg. The presence of the daughter isotope can be confirmed by another isotope anomaly observed in Ca, Al-rich inclusions. The ^{26}Mg/^{24}Mg ratio in Al-rich minerals is higher than normal, and the increase is correlated with the Al content. That the parent bodies of the chondrites were unmelted may be due to the fact that they agglomerated later, after the major part of the ^{26}Al had already decayed.

3.5 Regolith, Primordial Rare Gases, and the Solar Wind

On the surfaces of the small meteorite parent bodies many processes have taken place that have left evidence behind that is still recognizable in the chondrites. Due to impacts, an unconsolidated layer of rock fragments and dust, called a *regolith*, was formed that was turned over often and well mixed. (Such a regolith also covers the surface of the Moon.) Meteorite breccias were formed through the compaction of dust and debris in the regolith.

A special type of breccia has a *light-dark structure*: light-colored fragments enclosed in a darker matrix (Fig. 108). From the presence of the so-called *primordial rare gases* it can be demonstrated that the components of this breccia must have once resided in a regolith on the surface of a body without an atmosphere. One of the first meteorites in which these gases could be studied in detail was the Breitscheid chondrite that fell in 1956 in Germany. Measurements on Breitscheid samples at the Max-Planck-Institute for Chemistry in Mainz (with the apparatus shown in Fig. 99) showed unusually large amounts of noble gases and, in addition, that these amounts fluctuated widely from sample to sample. It was then discovered that these excess noble gases were contained only in the dark parts of the brecciated meteorite (Fig. 108). The content of helium was about 30 times and that of neon 9 times larger than normal, higher than could be explained from the decay of uranium and thorium, or as the result of cosmic rays. This noble gas component was then referred to as *primordial* because it was at first believed that it had a primordial origin, that it was

Fig. 108. Broken piece of the H–chondrite Breitscheid showing the light-dark structure, natural size. (After Hentschel, Geochim. Cosmochim. Acta 17, 1959)

present already in the solar nebula from which this particular material formed. Through a newly developed chemical etching technique the individual minerals of the meteorite could be measured separately, and the distribution of the noble gasses studied. It was revealed that all of the minerals contained the gases in a thin surface layer about 0.0001 mm thick. Thus, in 1965, H. Wänke proposed that the primordial gases originated from the *solar wind*, which would be trapped in the minerals of a regolith on the surface of a body without an atmosphere. A few years earlier the astrophysist, L. Bierman, proposed a solar wind on theoretical grounds in order to explain the deflection of comet trails as they passed the Sun. It is a corpuscular radiation that radiates from the Sun with an energy of around 10 MeV containing all of the elements in their solar abundances, i.e., predominately hydrogen and helium. The solar wind origin of the primordial gases was later confirmed by Moon landings, where the same primordial gases were found

Table 21. Element and isotope ratios in the solar wind

	He/Ne	^{3}He/^{4}He	^{20}Ne/^{22}Ne
Pantar metal phase	750	2200	14.0
Lunar experiment	540	2350	13.7

Source: Hintenberger et al., Z Naturforsch (1965); Geiss et al. (1972).

in the dust grains of the Moon. In Table 21 the abundances of primordial gases in meteorites are compared with solar wind noble gases captured directly on the Moon by Swiss investigators using a large aluminum foil. These measurements also permit important inferences regarding the processes in the outer layers of the Sun.

3.6 The Planets

Through collisions by the precursors of the planets (planetesimals), larger bodies were formed until finally the major planets that we know today evolved. The asteroids are for the most part probably the survivors of this process along with their debris. Very little is known about this process. The crater-saturated surfaces of the planets give evidence only of the final stages of this development. The investigation of asteroids and comets by space probes represents an important further step in our knowledge. In this category belongs the USSR 1990 mission to the Mars' moon Phobos. This irregularly shaped moon is probably a captured asteroid (Fig. 109). Nevertheless, the direct investigation of asteroidal fragments in the laboratory, i.e., the analysis of meteorites will remain indispensable.

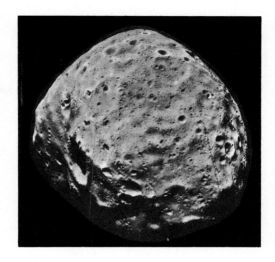

Fig. 109. The Mars moon Phobos, diameter 22–25 km, photographed by the Viking mars orbiter in 1977. (NASA photograph)

3.7 Conclusions

The investigation of meteorites, their different types, their minerals, and their trace elements and isotopes has provided us with a wealth of information about the Solar System. Because the chondrites contain primitive material that has changed very little, we can learn more about the earliest history of the Solar System. The interstellar grains which they contain can perhaps provide a view beyond the Solar System.

We have still not succeeded in completely understanding the details of the formation of the chondrules and chondrites. Figure 110 shows a section through the Tieschitz chondrite. Each chondrule that we see is an individual by itself, each has its own history. Taken

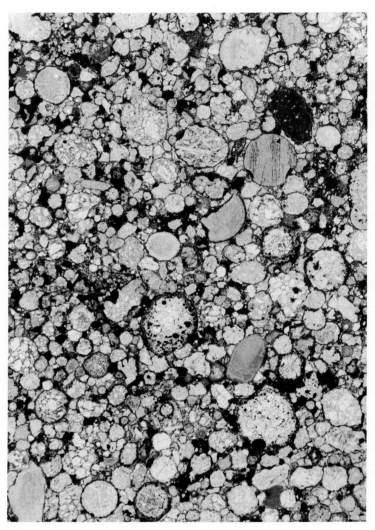

Fig. 110. Thin section of the H3 chondrite Tieschitz, about fivefold enlarged

together they have locked within them the story of the beginnings of our world, and we have still to find the key.

The scientific study of meteorites involves many disciplines, from astrophysics to nuclear physics, chemistry, and mineralogy. Nevertheless, the participation of laymen is both possible and necessary. Each newly observed or recovered meteorite may provide new understanding. The observations made during a meteorite fall, for instance, are essential and only possible through eye-witnesses who were accidently present. Small, seemingly insignificant details reported by laymen, if they are provided in sufficient number and recorded with sufficient precision, can form the background for broader and important conclusions. Therefore, the authors would consider it a satisfying reward, if many readers of this small book develop the readiness and ability to work together with us to further the science of meteoritics.

4 Appendix

4.1 Meteorite Collections and Research

As soon as the true nature of meteorites as extraterrestrial bodies was recognized, they were saved in collections. Several large collections were assembled already in the last century. These have been important for the study of meteorites, because the material is not readily available as terrestrial rocks are, and many meteorites are unique or of a rare type.

The oldest and still one of the most outstanding collections is that of the Museum of Natural History in Vienna. Similarly old and comprehensive are the collections of the British Museum (Natural History) in London and the Museum National d'Histoire Naturelle in Paris. In the USA there are three collections with more than 1000 meteorites in each, which were established already in the last century: in the National Museum of Natural History (Smithsonian Institution) in Washington, D.C., in the Field Museum in Chicago, and in the American Museum of Natural History in New York. Other large collections with about 500 meteorites are in the Institute of Meteoritics in Albuquerque, New Mexico, at the University of California Los Angeles, and at the Harvard University in Cambridge, Massachusetts. In Russia exists a Committee on Meteorites at the Academy of Sciences, which also has a large and important meteorite collection. More recent are the collections of Antarctic meteorites curated by the Polar Research Institute in Tokyo, Japan, and in the USA by NASA at the Johnson Space Center in Houston, Texas.

Active meteorite research is conducted today in all the Natural History Museums mentioned, as well as in other research institutions all over the world. These institutions will also answer questions about meteorites and are able to analyze stones or irons in order to determine whether they are real meteorites.

The following table lists such institutions in different countries:

Austria:
Naturhistorisches Museum, Mineralogisch-Petrologische Abteilung, P.O.Box 417, A-1014 Vienna.

Australia:
Western Australian Museum, Department of Mineralogy, Francis Street, Perth, Western Australia 6000.
Research School of Earth Sciences, Australian National University, Canberra, ACT 2600.

Czech Republic:
Narodni Muzeum V Praze, Vaclavski Namesti 68, 11579, Prague 1.

Denmark:
Dansk Tekniske Hogscole, Department of Metallurgy, Building 204, DK-2800 Lyngby.

England:
The Natural History Museum, Department of Mineralogy, Cromwell Road, London SW7 5BD, UK.

Department of Earth Sciences, The Open University, Milton Keynes MK7 6AA, UK.

France:
Museum d'Histoire Naturelle, Laboratoire de Minéralogie, 61 Rue Buffon, 75005 Paris.

Germany:
Max-Planck-Institut für Chemie, Abteilung Kosmochemie, Saarstrasse 23, D-55122 Mainz.
Institut für Planetologie, Universität Münster, Wilhelm–Klemm–Strasse 10, D-48149 Münster.
Museum für Naturkunde, Invalidenstrasse 43, D-10115 Berlin.

India:
Physical Research Laboratory, Ahmedabad 380009.
Department of Chemistry, Indian Institute of Technology, Kanpur 208016.

Italy:
Museo Civico di Storia Na., Corso Venezia 55, I-20121 Milano.
Sc. Terra, Univ. "La Sapienza", P. Le Aldo Moro 5, I-00185 Rome.

Japan:
National Institute of Polar Research, 9–10, Kaga 1-chome, Itabashiku, Tokyo 173.
Mineralogical Institute, Faculty of Science, University of Tokyo, Hongo, Tokyo 113.

Netherlands:
Dutch Meteor Society, Lederkarper 4, 2318 NB Leiden.
University of Utrecht, Van der Graaff Laboratory, P.O. Box 80.000, 3508 TA Utrecht.

Russia:
Committee on Meteorites, Russian Academy of

Sciences, Uliza M. Ulyanavoi 3, korp. 1, Moscow 117313.
Vernadsky Institute of Geochemistry and Analytical Chemistry, Russian Academy of Sciences, Ul. Kosigina 19, Moscow.

Switzerland:
Institut für Kristallographie u. Petrographie,
ETH Zürich, Soneggstrasse 5, CH-8092 Zürich.
Physikalisches Institut der Universität, Sidlerstrasse 5, CH-3012 Bern.

South Africa:
Department of Geochemistry, University of
Cape Town, Rondebosch, C.P.
B.P. Institute of Geophysics, University of the Witwatersrand, 2001 Johannesburg.

USA:
National Museum of Natural History, Department of Mineral Sciences, Smithsonian Institution,
Washington, D.C. 20560.
American Museum of Natural History, Department of Mineralogy, 79th Street at CPW, New York,
NY 10024.
Institute of Geophysics and Planetary Physics,
University of California, Los Angeles, CA 90024.
Institute of Meteoritics, University of New Mexico, Albuquerque, NM 87131.
Center for Meteorite Studies, Arizona State
University, Tempe, AZ 85287.
Planetary Geosciences Division, University of Hawaii, Honululu, HI 96822.
NASA Johnson Space Center, Code SN2, Houston, TX 77058.

4.2   Exchange Value of Meteorites

Meteorite collectors often exchange meteorite samples. Their exchange value depends on the mass of the specific meteorite fall or find and its type. The smaller the mass and the rarer the type, the higher its value.

The meteoriticist, E.A. Wülfing, tried to express this relation in a formula:

$$V = \sqrt[3]{\frac{1}{W \times T}} \, ,$$

where V is the relative value, W the weight of the meteorite fall or find, and T the total weight of its class. He chose the third root in order to obtain the relation between large and small meteorites in a reasonable range. In this way two meteorites differing in weight (or total weight of the class) by a factor of 1000 will differ in exchange value by a factor 10.

In Table 22 the class weights of the main meteorite classes (and their largest members) are given. From these values Fig. 111 was constructed which allows the direct determination of the exchange value of a given meteorite. Meteorite classes with a similar group weight are represented here by one line, as indicated in Table 22. In order to determine the relative value of a meteorite, one locates its weight on the lower axis (i.e., about 100 kg for the eucrite Juvinas), goes up from there until one reaches the line valid for its class (B for Juvinas), and from this intersection one can read on the left axis the relative value (20 for Juvinas). Similarly, one gets a value of 60 on line D for the H-chondrite Noblesville, a stone of about 0.5 kg which fell on August 31, 1991, in

Table 22. Weight and number of meteorites in different classes (excluding Antarctic meteorites)

	Class (number)	Total Weight (kg)	Largest member
Line A[a]	CI-chondrites (5)	17	Orgueil, 10 kg
	Ureilites (11)	20	Kenna, 11 kg
	SNC meteorites (6)	70	Nakhla, 40 kg
Line B	CM-chondrites (24)	150	Murchison, 100 kg
	Enstatite chondrites (23)	290	Abee, 107 kg
	C3O-chondrites (8)	290	Kainsaz, 200 kg
	Eucrites, howardites, diogenites (64)	380	Juvinas, 91 kg
Line C	Type 3 ordinary chondrites (84)	800	Clovis No.1, 283 kg
	Aubrites (10)	1100	Norton Co., 1000 kg
	LL-chondrites (87)	1400	Paragould, 400 kg
	C3V-chondrites (12)	2050	Allende, 2 t
	Mesosiderites (29)	2200	Bondoc, 890 kg
Line D	Hexahedrites (49)	4400	Coahuila, 2 t
	Pallasites (41)	7800	Huckitta, 2 t
	Ataxites (33)	9500[b]	Sta. Catharina, 7 t
	L-chondrites, types 4–6 (705)	10300	Long Island, 560 kg
	H-chondrites, types 4–6 (737)	11500	Jilin, 4 t
Line E	Octahedrites (485)	235000	Cape York, 58 t

[a] Lines A–E refer to Fig. 110.
[b] Without Hoba, 60 t.

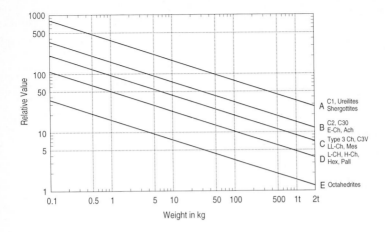

Fig. 111. Exchange value of meteorites. The *lines* A through E correspond to the meteorite classes given in Table 22

Indiana, USA. Thus, Noblesville would be three times as valuable as Juvinas, or for 1 g of the former one could ask for 3 g of the latter.

The values of a meteorite also depend on other factors, however. Observed falls are usually regarded as more valuable than finds, especially if these are weathered. Collectors like complete specimens with a fusion crust better than cut slices. Certain meteorite types are more interesting for scientific study than others; for example, the primitive carbonaceous chondrites compared to the ordinary chondrites. But also certain meteorites of a common type can be quite valuable for scientific study, like the ordinary chondrites containing primordial rare gases (see p. 196). Thus, Wülfing's formula and Fig. 111 should not be taken too seriously; each meteorite collector and investigator will have his own scale of values.

4.3 Etching of Iron Meteorites

In order to reveal the Widmanstätten pattern iron meteorites have to be etched. First, a smooth surface is produced by grinding and polishing. The specimen is put into a flat bowl and the etchant is applied with a soft brush. As an etchant Nital (a 5% solution of nitric acid in alcohol) is most satisfactory. An aqueous solution of the acid may be used, if the iron is free of troilite inclusions, because it develops a brown discoloration around such inclusions. The etchant should be spread evenly over the surface and kept constantly in motion, until the desired distinctness of structure is developed. Very thorough washing under a stream of warm water is then necessary to avoid rusting. The specimen should be dried in hot air and preferably kept warm for some time. The surface may then be protected by covering it with a thin layer of lacquer.

4.4 Detection of Nickel

All iron meteorites contain nickel and the detection of this element in a metal piece may serve to distinguish meteorites from artificial irons. The detection of nickel may be done by a color reaction.

First, a small portion of the metal has to be dissolved in acid. This may be achieved by placing a drop of dilute hydrochloric acid (10%) on the metal surface. After some time the acid will become yellow by the dissolved iron. It can now be removed with a pipette and placed on a white porcellain plate. Here, the acid is neutralized with a drop of dilute ammonia. The proof of nickel is accomplished with dimethylglyoxime. A drop of a 1% solution of this reagent in alcohol is applied to the test solution. If nickel is present, a red color or precipitation develops.

4.5 Meteorite Falls from 1985 Through 1992

(Source: Meteoritical Bulletin, No. 65–68, Editor A.L. Graham, and No. 69–75, Editor F. Wlotzka)

1. La Criolla   31°14'S, 58°10'W
15 miles WNW of Concordia, Entre Rios, Argentina
Fell on January 6, 1985. L6 chondrite.
After a bright fireball and many detonations, tens of crusted stones fell over a 7 × 10 km ellipsoid area east of Estancion La Criolla. The largest stones recovered weighed 6.1 kg, 1.95 kg, and 750 g, the total mass found was over 35 kg.

2. Salzwedel   52°48'N, 11°12'E
Magdeburg, Germany
Fell on November 14, 1985, 18:17 h. L6 chondrite.
A fireball of about half the size of the moon was seen over Lower Saxony between Hannover and Braunschweig; it was moving from SW to NE. A bang, similar to a detonation, was heard about 2 to 3 min later near the NE end of the path. At about the same time, in the village Hohenlangenbeck near Salzwedel, a falling object was seen hitting a tree. On the next day, a boy found the meteorite, a single mass weighing 43 g.

3. Tianzhang   32°56'48"N, 118°59'24"E
Gaomiao, Tianzhang County, Anhui, China
Fell on January 28, 1986, 17:00 h. H5 chondrite.
A single mass weighing 2232 g was recovered from a field by a farmer a few minutes after the fall.

4. Suizhou   31°43'N, 113°23'E
Suizhou County, Hubei, China
Fell on April 15, 1986, 18:50 h. L6 chondrite.
Approximately 12 specimens having a total weight of about 70 kg, were recovered.

5. Lanxi   46°14'30"N, 126°11'46"E
Hongxing, Lanxi County, Heilongjiang, China
Fell on June 10, 1986, 10:00 h. L6 chondrite.
A single mass weighing 1282 g was recovered by a farmer from a flax field about 20 min after the fall.

6. Kokubunji   34°18'N, 133°57'E
Kokubunji-cho, Kagawa prefecture, Shikoku, Japan
Fell on July 29, 1986, 19:00 h. L6 chondrite.
After a fireball and detonations many stones fell, the largest weighing about 10 kg. Many stones struck tiled roofs or paved roads. The total mass recovered was about 11 kg.

7. Wuan   36°45'N, 114°15'E
Wuan County, Hebei Province, China
Fell on July 31, 1986, 11:00 h. H6 chondrite.
A single 50-kg mass fell in Wuan County, after detonations were heard.

8. Raghunathpura   27°43'31"N, 76°27'54"E
Alwar district, Rajasthan, India
Fell on November 20, 1986, about 20:00 h. Iron, hexahedrite (IIA).

The meteorite fall was observed and a single mass of 10 kg was recovered the next day from a meter deep pit.

9. Laochenzhen   33°08'N, 115°10'E
Shenqiu County, Henan Province, China
Fell on February 23, 1987, 03:00 h. H5 chondrite.
After detonations one mass of 14.25 kg fell in Yangwa suburb of Laochenzhen town.

10. Greenwell Springs   30° 30'55"N, 91° 00'44"W
Baton Rouge, Louisiana, USA
Fell within a week before November 30, 1987. LL4 chondrite.
A single stone of 664 g was found in the frontyard of a house. The fall must have occurred within 1 week before the recovery.

11. Chisenga   10°3'34"S, 33°23'42"E
Chitipa District, Malawi
Fell on January 17, 1988. Iron. Coarse octahedrite(1.5-mm kamacite bandwidth).
The fall was eyewitnessed by a woman who was 12.5 m away from the landing site, it was accompanied by a loud explosion. A mass of 3.92 kg was recovered by the police from a hole 30 cm deep. The landing site is about 50 km from Chitipa Boma, for which the coordinates are given. The main mass is at the National Museum of Malawi.

12. Trebbin   53°13'N, 13°10'E
Potsdam district, near Berlin, Germany
Fell on March 1, 1988, 13:30 h. LL6 chondrite.
The fall was accompanied by a loud hissing noise and, later, by a detonation-like clap. Because of the clouded sky, no light phenomena were seen. The meteorite hit

the glass roof of a greenhouse and broke into numerous fragments. The total mass is 1250 g.

13. Torino   45°03'N, 07°45'E
Torino, Italy
Fell on May 18, 1988, 13:40 h. H6 chondrite.
A shower of stones fell during a thunderstorm in the city of Torino. The largest piece of about 800 g fell in a parking lot, about a foot away from a parked car. The stone made a 15-cm-wide, 3-cm-deep pit on the tar road. Several smaller fragments were later found in suburbs of Torino, the total recovered weight being about 1 kg. Measurement of cosmic ray-produced rare gases yielded an exposure age of 48 million years.

14. Chela   3°40'S, 32°30'E
Kahama district, Tanzania
Fell on July 12, 1988, 11:40 h. H5 chondrite.
Several stones fell after loud noise and detonations. They were collected by local people and two masses of 1 and 1.8 kg were handed to the police.

15. Ceniceros   26°28'N, 105°14'W
Chihuahua, Mexico
Fell on August 20, 1988, 10:20 h. H3 chondrite.
A single stone of 1 kg fell in a field and was found on the same day by a farmer.

16. Uchkuduk   41°46'N, 62°31'E
Bukharskaya region, Uzbekistan
Fell on June 21, 1989, 18:00 h. L6 chondrite.
Cowboys heard a sonic boom and saw a small dust cloud at the impact site. They found two fragments totaling about 1 kg in a small crater in a sand field.

17. Sixiangkou   32°26'N, 119°52'E
Jiangsu, China

Fell on August 15, 1989, 21:53 h. L6 chondrite, veined.

After a sonic boom was heard, a meteorite hit the roof of a house. Four stones, total weight 630 g, were recovered from the roof, a vegetable field, and the roadside.

18. Sfax   34°45'N, 10°43'E
Tunisia
Fell on October 16, 1989, 09:30 h. L6 chondrite.

Following an explosion, a meteorite fell about 10 km north of Sfax. At least four fragments (weighing 4.2 kg, 500 g, and two others) were recovered.

19. Glanerbrug   52°13'N, 6°57'E
Near Enschede, Netherlands
Fell on April 7, 1990, 20:34 h. LL chondrite breccia.

A stony meteorite fell on the roof of a house. It broke into numerous fragments, the largest weighing 135 g. The total recovered mass was about 500 g. A fireball was reported by several hundred people in The Netherlands and Germany, orbital elements were calculated from these observations. The meteorite is now at the National Museum of Natural History, Leiden, The Netherlands.

20. Sterlitamak   53°40'N, 55°59'E
Bashkir ASSR, Russia
Fell on May 17, 1990. Iron. Medium octahedrite.

Fragments totaling about 10 kg (the largest 6.6 kg) were recovered from impact craters. The main mass is still in the crater.

21. Magombedze   19°29'S, 31°39'E
Masvingo District, Zimbabwe
Fell on July 2, 1990, 17:30 h. H6 chondrite, brecciated.

Several stones are reported to have fallen after three separate detonations were heard near the village Mogambedze, 54 km ENE of the town Gutu, for which the coordinates are given. Three larger fragments and several small chips having a total weight of 666.6 g were recovered.

22. Tahara   34°43'N, 137°18'E
Tahara-machi, Japan
Fell on March 26, 1991. H5 chondrite.
The meteorite fell on to the deck of a ship which was loading cars in Toyohashi harbor. The crew found the stony fragments when they returned from lunch after 12:00. Most of them were swept into the water during deck cleaning, but about 1 kg was preserved. No sound or light was observed before discovering the meteorite.

23. Glatton   52°27'35"N, 0°18'0"W
Cambridgeshire, England
Fell on May 5, 1991, 12:30 local time. L6 chondrite.
A single mass weighing 767 g was observed to fall by a man working in his garden. No light or sound was noticed. Despite a search in the area, no further material was recovered.

24. Noblesville   40°5'7"N, 86°3'18"W
Hamilton County, Indiana, USA
Fell on August 31, 1991, at about 19:00 CDT.
H chondrite.
The stone passed two boys who observed it land 3.50 m in front of them on the lawn before a house. No light or sound except for the whirring sound as it passed and the thud as it hit the ground were noticed. It is an oriented specimen with well-developed flight markings, weight 483.7 g. Large white H6-clasts are embedded in an H4 host.

25. Mbale   1°04'N, 34°10'E
Uganda
Fell on August 14, 1992. L5 chondrite.
A loud explosion was heard over the town Mbale, and later a smoke trail was seen in the sky. A meteorite shower fell over an area of about 3 × 7 km. More than 400 stones were recovered, the largest weighing 24 kg. The stones hit several buildings, but nobody was hurt. The total known weight is 110 kg, but it is estimated that local people recovered an additional 200 to 400 kg.

26. Peekskill   41°17'N, 73°55'W
New York, USA
Fell on October 9, 1992, 19:50 h. H6 chondrite.
A brilliant fireball was seen in the eastern United States between Virginia and New Jersey, moving northeast. A stone of 12.4 kg fell and damaged a parked car, penetrating the trunk in the rear.

27. Mihonoseki   35°34'N, 133°13'E
Yatsuka-gun, Shimane-ken, Japan
Fell on December 10, 1992, 21:00 h. L6 chondrite.
A fireball was observed in Wakayama, Osaka, Hyogo, Okayama, and Hiroshima prefectures and on Shikoku Island. The meteorite fell through the roof of a house and penetrated two floors. A 6.4-kg stone was found the next day under the house next to a small hole in the ground. Nobody was hurt inside the house.

4.6 Literature

The journal *Meteoritics* is published by the Meteoritical Society, an association of meteorite collectors, scientists, and friends of meteoritics. The journal prints articles about meteorites, meteors, lunar samples, impact craters, interplanetary dust, tektites, related aspects of asteroids, comets, and planets, and on the origin and history of the Solar System. Part of the journal is the *Meteoritical Bulletin* which publishes data on falls or finds and classification of all new meteorites. Members of the Society receive this journal for a fee of $ 45 (1993). Membership applications may be sent to the Treasurer: Professor Dr. Joseph I. Goldstein, College of Engineering, University of Massachusetts, Amherst, MA 01003, USA

All known meteorites up to 1985 are listed in the *Catalogue of Meteorites* by A.L. Graham, A.W.R. Bevan, and R. Hutchison, published by the British Museum (Natural History), London 1985.

The older literature up to 1950 about meteorites is compiled in *A Bibliography on Meteorites* by H. Brown, G. Kullerud, and W. Nichiporuk, University of Chicago Press, 1953.

The following is a list of recent books on meteorites:

Buchwald VF (1975) Handbook of iron meteorites. Their history, distribution, composition and structure, 3 vols., University of California Press, Berkeley

Buchwald VF (1992) Meteoritter – Nöglen til Jordens fortid. Gyldendal, Oslo (in Danish)

Burke JT (1986) Cosmic debris – meteorites in history. University of California Press, Berkeley

Dodd RT (1981) Meteorites, a petrologic-chemical synthesis. Cambridge University Press, Cambridge

Dodd RT (1986) Thunderstones and shooting stars. Harvard University Press, Cambridge

Kerridge JF, Matthews MS (eds) (1988) Meteorites and the early Solar System. The University of Arizona Press, Tucson

Mark K (1987) Meteorite craters. The University of Arizona Press, Tucson

Mason B (1962) Meteorites. Wiley, New York

Mason B (ed) (1971) Handbook of elemental abundances in meteorites. Gordon and Breach, New York

McSween HY (1986) Meteorites and their parent planets. Cambridge University Press, Cambridge

Nininger HH (1956) Arizona's meteorite crater. Past - present - future. American Meteorite Laboratory, Denver

Nininger HH (1972) Find a falling star. PS Ericson, New York

O'Keefe JAO (1976) Tektites and their origin. Elsevier, Amsterdam

Wasson JT (1974) Meteorites. Springer, Berlin Heidelberg New York

Wasson JT (1985) Meteorites – their record of early Solar-System history. Freeman, New York

Table 23. Composition of meteorites and the earth's crust (values in ppm, except where % is indicated)

Number		Element	CI-chondrites	H-chondrites	Eucrite	Upper Earth's crust
3	Li	Lithium	1.49	1.7+	6.1+	20
4	Be	Beryllium	0.025	0.04	0.04	
5	B	Boron	0.87	1.0+	0.83+	
6	C	Carbon	3.39%	800	600	320
7	N	Nitrogen	3180	50+	40	20
8	O	Oxygen	44.8%	35.1%	42.4%	47.3%
9	F	Fluorine	58.2	60+	19	720
11	Na	Sodium	0.50%	0.71%	0.28%	2.45%
12	Mg	Magnesium	9.26%	13.85%	4.0%	1.39%
13	Al	Aluminum	0.86%	1.05%	7.1%	7.83%
14	Si	Silicon	10.26%	16.3%	23.0%	30.54%
15	P	Phosphorus	1200	1040	400	810
16	S	Sulphur	5.53%	1.42%	0.20%	310
17	Cl	Chlorine	698	100	18	320
19	K	Potassium	544	720	222	2.82%
20	Ca	Calcium	0.95%	1.15%	7.7%	2.87%
21	Sc	Scandium	5.9	9.8	28.5	14
22	Ti	Titanium	476	600	3800	4700
23	V	Vanadium	55.0	61+	75+	95
24	Cr	Chromium	2646	3200	2090	70
25	Mn	Manganese	1938	2300	3990	690
26	Fe	Iron	18.23%	29.0%	14.5%	3.53%
27	Co	Cobalt	506	811	5.8	12
28	Ni	Nickel	1.077%	1.72%	1.1	44
29	Cu	Copper	131	90	1.7	24
30	Zn	Zink	323	51	1.1	57
31	Ga	Gallium	9.71	5+	2.6	17
32	Ge	Germanium	32.7	7	0.004	1.3
33	As	Arsenium	1.81	2.1+	0.18	1.7
34	Se	Selenium	20.7	7.7	0.077	0.09
35	Br	Bromine	3.50	0.2+	0.16	1.0

Table 23. Continued

Number		Element	CI-chondrites	H-chondrites	Eucrite	Upper Earth's crust
37	Rb	Rubidium	2.32	2.54	0.25	120
38	Sr	Strontium	7.26	8.23	78	290
39	Y	Yttrium	1.57	2.2+	16	30
40	Zr	Zirkonium	3.87	7.24	46	140
41	Nb	Niobium	0.246	<0.2+	2.7	14
42	Mo	Molybdenum	0.93	1.7+		
44	Ru	Ruthenium	0.712	0.82		
45	Rh	Rhodium	0.13	0.2+		
46	Pd	Palladium	0.556	1.1+	0.0004	
47	Ag	Silver	0.197	0.05	0.1	0.06
48	Cd	Cadmium	0.68	0.02	0.03	0.1
49	In	Indium	0.0778	0.00025	0.0001	0.07
50	Sn	Tin	1.68	0.46		
51	Sb	Antimony	0.133	0.09	0.042	0.2
52	Te	Tellurium	2.27	0.74	0.00001	0.002
53	J	Iodine	0.433	0.04	0.2	0.5
55	Cs	Caesium	0.188	0.08	0.005	2.7
56	Ba	Barium	2.41	4.0	31	730
57	La	Lanthanum	0.25	0.32	2.6	44
58	Ce	Cerium	0.64	0.48		75
59	Pr	Praseodymium	0.096	0.12	0.94	7.6
60	Nd	Neodymium	0.474	0.61	5.1	30
62	Sm	Samarium	0.154	0.20	1.5	6.6
63	Eu	Europium	0.058	0.081	0.61	1.4
64	Gd	Gadolinium	0.204	0.34	2.3	8
65	Tb	Terbium	0.037	0.053	0.60	1.4
66	Dy	Dysprosium	0.254	0.34	3.2	6.1
67	Ho	Holmium	0.057	0.068	0.42	1.8
68	Er	Erbium	0.166	0.205	2.3	3.4
69	Tm	Thulium	0.026	0.033		
70	Yb	Ytterbium	0.165	0.19	1.72	3.4
71	Lu	Lutetium	0.025	0.033	0.28	0.6
72	Hf	Hafnium	0.107	0.204	1.3	3
73	Ta	Tantal	0.014		0.12	3.4
74	W	Tungsten	0.095	0.13	0.041	1.3
75	Re	Rhenium	0.0383	0.081	0.00001	0.001
76	Os	Osmium	0.486	0.814	0.00002	0.001

Table 23. Continued

Number		Element	CI-chondrites	H-	Eucrite	Upper Earth's crust
77	Ir	Iridium	0.459	0.77	0.00003	0.001
78	Pt	Platinum	0.994	1.7		0.005
79	Au	Gold	0.152	0.25	0.007	0.004
80	Hg	Mercury	0.258			
81	Tl	Thallium	0.143	0.0025	0.001	0.02
82	Pb	Lead	2.47	(0.2)	0.4	0.46
83	Bi	Bismuth	0.111	0.0014	0.0035	0.01
90	Th	Thorium	0.0286	0.038	0.06+	11
92	U	Uranium	0.0082	0.011	0.09	3.5

Values for CI-chondrites from Palme and Beer (1993); for H-chondrite
(Richardton) and eucrite (Juvinas) from Palme, Suess, and Zeh, Landolt-
Börnstein, Neue Serie VI/2a, Springer, Berlin Heidelberg New York 1981.
Values with + from Mason, Handbook of Elemental Abundances in Meteorites,
Gordon and Breach 1971.
Values for the upper Earth crust from Wedepohl, Fortschr Miner 52, 1975.

5 Subject Index

229

Springer-Verlag
and the Environment

We at Springer-Verlag firmly believe that an international science publisher has a special obligation to the environment, and our corporate policies consistently reflect this conviction.

We also expect our business partners – paper mills, printers, packaging manufacturers, etc. – to commit themselves to using environmentally friendly materials and production processes.

The paper in this book is made from low- or no-chlorine pulp and is acid free, in conformance with international standards for paper permanency.